ORGANIZATIONAL HEALTH IN A CLIMATE OF FEAR AND UNCERTAINTY

REVITALIZING ORGANIZATIONS THROUGH PROMOTING HUMAN VALUES

ORGANIZATIONAL HEALTH IN A CLIMATE OF FEAR AND UNCERTAINTY

REVITALIZING ORGANIZATIONS THROUGH PROMOTING HUMAN VALUES

Sheila Keegan

EMERGENT™
PUBLICATIONS

Organizational Health in a Climate of Fear and Uncertainty
Revitalizing Organizations through Promoting Human Values
Written by: Sheila Keegan

Library of Congress Control Number: 2012954011

ISBN: 978-1-938158-03-2

Copyright © 2012
Emergent Publications,
3810 N. 188th Ave, Litchfield Park, AZ 85340, USA

Printed in the United States of America

CONTENTS

ACKNOWLEDGEMENTS

With very many thanks to Rosie Campbell, my long term business partner who, with patience and good humour, has dutifully read and commented on various drafts of this essay.

Thank you also to Dr. Mike Short CBE, Vice President of Telefonica Europe and to Dr. Karen Norman, Research Supervisor, Complexity and Management Group, Hertfordshire Business School, both of whom generously agreed to read this essay and made constructive comments and suggestions.

ABOUT THE AUTHOR

Dr. **Sheila Keegan** is a business psychologist, qualitative researcher and organizational change consultant. She is a founding partner of Campbell Keegan Ltd, a consultancy that works with private and public sector clients, within the broad areas of change and communications; this includes facilitating new thinking and strategic development related to brands, products, services, internal communications and organizational culture.

Her main interests lie in introducing ways of thinking and practice from the complexity and behavioral sciences into organizational life and also 're-humanizing' the workplace, for example by encouraging more improvisational and 'joined up' thinking as a counterbalance to 'over-structuring' and target culture.

Sheila is also an author, a trainer in qualitative research approaches, a Master Practitioner in NLP, a Fellow of the Market Research Society and an Associate Fellow of the British Psychological Society. She is a regular speaker at conferences, universities and to government and commercial bodies and acts as an advisor on cultural trends and consumer psychology to companies, advertising agencies, think-tanks, public service companies and the media.

INTRODUCTION

We live in strange times. Who would have thought, five years ago, that international banks would crash, triggering a collapse in economies worldwide; that we would be plunged into the worst recession since the 1930s, a recession that would force us to re-evaluate structures, systems, life-styles and assumptions that once seemed invincible? Who would have guessed that we, as a society, would be questioning the whole capitalist dream, contemplating the hollowness of obsessive materialism, wallowing in a mood of dis-ease and uncertainty?

Within the UK, Public Sector finance has been slashed. As I write, the European Union is tottering on the brink of collapse and EU citizens are taking refuge in London as economic immigrants. Meanwhile, UK bankers are receiving bonuses that would support minor fiefdoms, even as banking scandals reverberate throughout the globe. Arguably, the British people have never felt so disillusioned, uncertain and dispirited about their futures.

Organizational life has, inevitably, felt the knock on effect of the downturn and the unease that accompanies it. Some companies that were household names have disappeared. Redundancy, high unemployment and lower wages are common. In addition, management fashions, over the last couple of decades, have moved towards increased performance monitoring; target setting for individuals, departments and boards that are increasingly beholden to shareholders. At best these targets can provide useful steers. At worst, they can be lethal. In many situations they undermine morale and employees' sense of self-worth.

Understanding how British consumers view the current recession as it is emerging; how they feel about working in recession-hit organizations and how this affects their consumption, their moods and levels of optimism, is integral to the future health of consumer-facing—indeed all—organizations.

How can we start to fight back from the gloom and despondency that has gripped our nation—and beyond? Can we restore our sense of autonomy, pride in our work and an increase in our productivity? Can organizational malaise be reverse, when fear and uncertainty have pervaded the working environments of so many of us over the last four years?

There are no easy answers. This essay explores some of these issues within the context of working life in a range of UK organizations. It is based, in large part, on research projects carried out by the business psychology consultancy, Campbell Keegan Ltd., within large organizations, both in the private and public sectors, over the last 20 or more years. It includes input from books, papers, conferences and conversations with colleagues, particularly academics and practitioners working in the area of organizational change.

My aim in writing this essay is to encourage thoughts, comments and discussion amongst those who work with, and within, organizations in the UK and other areas of the world, so that we may share experiences and further develop ideas on the nature of organizations in this time of anxiety and austerity. My initial aim was to direct this essay towards research practitioners working within organizations. However, as I wrote, I came increasingly to the view that, as almost all of us work in or have contact with organizations, it may have broader relevance. This essay draws on both the academic and the practical. It includes theory and case studies, although the main theoretical input comes from the complexity sciences and, in particular, complex responsive processes of relating (Stacey, 2001: 4-7, 2003: 51-54). This essay also touches on 'Eastern' and 'Western' thinking. Overall, it encourages the development of *Qualitative Mind* and *Qualitative Productivity* as ways of working within organizations, with the aim of integrating qualitative and quantitative perspectives and viewing 'the organization itself' as a process of complex responsive processes of relating.

THE SHAPING OF SOCIETY AND ORGANIZATIONAL LIFE

In 2008, the current world-wide financial crisis erupted, as if from no-where, and swept quickly around much of the developed world like a tornado; unexpected and unwelcome, a bolt out of the blue. At least that is how it was experienced by the majority of ordinary UK citizens who did not have access to knowledge that the major banks and the government may have possessed. Initially we all hoped it would be a blip, a brief period of belt-tightening and then it would be back to business as usual. However, this has not happened. Instead we slipped relentlessly into recession. Months passed into years and the belt-tightening has intensified. For many of us in the UK, financial worries and uncertainty have become part and parcel of the fabric of our lives; a culture of thrift has become established, as we hoard coupons, seek out discounts as a matter of course and cut back on indulgences.

Confidence in our society and our future has become undermined and we are left with a residue of anxiety and helplessness, even if we are not directly affected by redundancy, cutbacks or wage freezes. The general public is baffled that financial structures and established institutions that we grew up with—assumed pillars of society—are collapsing, no longer providing the comfortable familiarity and context for our lives. Financial institutions in general, city bonuses and the recent libor rate scandal have become the butt of consumer anger. While the general population skimps and saves, the bankers are seen as 'fat cats', the recipients of huge bonuses in spite of poor performances (Campbell Keegan, 2008/2011).

Economists describe these changes in a different language, but their sentiments are similar. The late-2000s financial crisis is considered by many economists to be the worst financial crisis since the Great Depression of the 1930s. It has resulted in the collapse of large financial institutions, the bailout of banks by national governments and downturns in stock markets around the world. Housing markets have also suffered, resulting in numerous evictions, foreclosures and prolonged unemployment. It contributed to the failure of key businesses, declines in consumer wealth estimated in the trillions of U.S. dollars, and a significant decline in economic activity, which lead to the severe global economic recession in 2008.

The financial crisis was triggered by a complex interplay of problems within the international banking systems. The bursting of the U.S. housing bubble, which peaked in 2007, caused the values of securities tied to U.S. real estate pricing to plummet, damaging financial institutions globally. Questions regarding bank solvency, declines in credit availability and damaged investor confidence had an impact on global stock markets, where securities suffered large losses during 2008 and early 2009. Economies worldwide slowed during this period, as credit tightened and international trade declined. Governments and central banks responded with unprecedented fiscal stimulus, monetary policy expansion and institutional bailouts.

Many causes for the financial crisis have been suggested. The United States Senate issued the Levin–Coburn Report, which found "that the crisis was not a natural disaster, but the result of high risk, complex financial products; undisclosed conflicts of interest; and the failure of regulators, the credit rating agencies, and the market itself to rein in the excesses of Wall Street.

Critics argued that credit rating agencies and investors failed to accurately price the risk involved with mortgage-related financial products, and that governments did not adjust their regulatory practices to address 21st-century financial markets. The 1999 repeal of the Glass–Steagall Act of 1933 effectively removed the separation that previously existed between Wall Street investment banks and depository banks. In response to the financial crisis, both market-based and regulatory solutions have been implemented or are under consideration.

Of course the crisis did not stop with the financial institutions. It has been argued that a range of other factors contributed to the climate of fear and uncertainty; the rising costs of consumption and falling house prices, competition from lower cost countries and/or moving manufacturing to the East, the polarizing effect of digital technologies, the changing roles of public/private partnerships and what this entails for employees.

These factors have combined to create a knock-on effect throughout the whole of society. Within the UK, unemployment, amongst the young has risen to 22% (June 2012 Government figures). Those who are in work are often uncertain about their future employment prospects. Within all manner of organizations, both large and small, there is considerable unease. Uncertainty breeds fear. Both the public and private sectors have experienced savages cuts in staffing levels and most organizations expect further cuts or restructuring. How can organizations survive, let along thrive, in this gloomy climate? What can senior management do to motivate staff; to instil a sense of purpose and productivity within this ever-changing environment? If we cannot predict the future of our organizations with any degree of certainty, how can we plan? Can managers instil confidence and enthusiasm in their staff when redundancy is the elephant in the room? Risk taking is the corner-stone of innovative organizations, but who wants to take risks in this climate?

'The organization' as on-going processes of relating

We take it for granted that in order to understand the culture of a particular organization; its decision-making processes and what makes it tick, we need to understand its history, its structure, its stated goals, what sector it is operating within, its profit margins and the myriad of other financial and sector data that define it. These factors are important but in themselves they do not fully *define* the organization. We also need to acknowledge the simultaneous on-going and changing human relationships and patterning that *is* 'an organization'; shaped by the continually evolving ethos, norms and values that constitute 'the organization'. From this perspective, 'an organization' is a complex, evolving, web of relationships that transcends the different levels and functions within that organization. It assumes that how people work, think and create together *is the organization*. Consequentially, the emerging culture, to a large extent, is what differentiates one organization from similar organizations.

While it is tempting to view organizations as isolated entities, existing within their own space and playing by their own rules, in reality organizations are viral networks, connecting prevailing cultures and sub-cultures. They continually influence and are influenced by the wider world outside the organization. No organization is an island and nothing happens in a vacuum. So to begin to understand what makes a particular organization *tick*, we need to understand how that organization is situated within its particular time, place and culture, and the factors that are shaping it.

Just as individuals are connected in myriad ways through genetic links, relationships, cultures, interests, friendships and much more, so too are organizations.

Many of us are familiar with the interesting—though difficult to prove—theory of *six degrees of separation*. This idea, proposed by Frigyes Karinthy, the Hungarian playwright, and popularized in a play written by John Guare, suggests that every individual is, on average, approximately six steps removed, by way of introduction, from any other person on earth. Therefore, a chain of "friend of a friend" connections can be made, on average, to link any two people on the planet in six steps or fewer. A similar principle might be applied to organizations, given that they are composed of individuals, each a hypothetical six steps or less away from every other individual. Organizations are connected with other organizations, either directly through the people who work there, and/or through shared values, interests and so on. This goes some way to explaining why similar ideas crop up at the same time in very different contexts or continents. We are constantly being influenced; shaped both consciously and unconsciously by our environment and, to a large extent, we are completely unaware of these influences on our attitudes, our beliefs and our behaviors.

In his fascinating book, *Thinking, Fast and Slow*, the Nobel Laureate in Economics, Daniel Kahneman, describes an elegantly simple survey he carried out with German students which illustrates how easily influenced we are as individuals and the ease with which our ideas and beliefs can be shaped, often without our conscious awareness. (Kahneman, 2011: 101). The students were asked two identical questions. The only difference was that group A was asked the questions in one order and group B was asked them in the reverse order. The two questions were:

1. How happy are you these days?
2. How many dates did you have last month?

The experimenters were interested in the correlation between the two answers. Would students who reported many dates claim to be happier than those with fewer dates? The results were surprising. When students were asked about their happiness levels first, followed by the number of dates, there was almost no correlation between number of dates and claimed happiness. However, when the questions were reversed, so that students were first asked about number of dates, followed by their reported level of happiness, the correlation was "about as high as correlations between psychological measures can get" (Kahneman, 2011: 102). Kahneman explains this difference as follows: Students who had many dates were reminded of a happy aspect of their life, whilst those who had none were reminded of loneliness and rejection—and this emotive frame of mind influenced their response to the question, 'How happy are you these days? If Kahneman's hypothesis is correct, then it seems as if asking students about their romantic life *primed* them to frame their subsequent answer in a particular way which reflected the question order. This is just one of many such intriguing studies that Kahneman describes in his book.

It can be quite disconcerting to realize that beliefs and values which we consider to be individual and deep-seated, can so easily be 'adjusted' and that emotions are so contagious that they can spread like wildfire, drawing individuals into their thrall.

The well documented riots that took place in London in the summer of 2011 sent shockwaves around the world, not least because the disturbances were broadcast in 'real time' on national television. Viewers watched large swathes of London being ransacked and set on fire. Many of the participants in the riots were interviewed, both during the course of the rioting itself and in subsequent days. There was a striking similarity in some of the explanations that rioters gave for their behavior. Many looked puzzled when asked why they did it. They did not seem to know. They just 'got caught up in it'. The micro-climate of rioting and the scores of people 'doing it', encouraged them to join in, they claimed. Of course this is not a justification, but it is an interesting phenomenon, perhaps reflecting the need to belong, to be part of, or to succumb to the emotional pull of 'acting as one'. If nothing else, it illustrates the power of contagion; how we can act in ways that undermine our normal values when caught up in an overwhelmingly emotive situation.

Recent research in neuroscience is beginning to explain this power of 'contagion' which is brought about, in part, by 'mirror neurons' in the brain which help us to understand the actions of others, and prime us to imitate what we see. (Neuroscientist Vilayanur Ramachandran, www.ted.com/talks). It is quite disconcerting to realize that, for all of us, our beliefs, emotions, attitudes and behaviors are more whimsical and context-driven than we would like to believe.

How priming and framing shape our perceptions

It is not only extreme situations that encourage us to mirror the behavior of others. *Priming*—an increased sensitivity to particular stimuli as a result of previous experience—prepares us to interpret the external world in certain culturally defined ways; for example expecting an exam to be stressful. The related concept of *framing* describes the way in which our perceptions can be altered by background or context. For example, asking, "Do you *really* want another cake?" is clearly worded in such a way as to shape a particular response. Priming and framing are crucial and on-going influences on our perceptions. Our behaviors shape and are shaped, in large part, by other people, often without us having any idea that it is happening, because such influences operate below our conscious awareness, in our adaptive unconscious (Wilson, 2002: 17-41).

How does all this relate to organizational life? If we accept that an organization is a network of on-going relationships, then it follows that organizations are susceptible to the same processes of group priming, framing, and emotional contagion as are individuals. Whilst we like to maintain the illusion that we effect change primarily through our conscious behaviors, the reality is that we are often unaware of our motivations and what triggers our actions. So we may be surprised by unexpected events or even our own behavior which seems illogical or out of character when we consider them in retrospect. We had been *primed* to expect another outcome or maybe we had adopted an inappropriate *frame*.

At present a key factor that *frames* and *primes* organizational life is the economic downturn and its protracted and widespread ramifications. This essay explores some of the ways in which the recession is impacting on, and shaping, organizational life at a human level, looking in particular at the effects of management initiatives, for example on relationship networks within organizations, on quantitative performance targets and on the cult of blame. Equally we will look at the flip side; how we might encourage the development of healthier, more productive organizations, through integrating qualitative and quantitative approaches, acknowledging the importance of emotion in decision making, absorbing contributions from neuroscience and generally promoting the human aspects of organizational life.

My hope is that the post-recession era will trigger a reassessment of the way in which organizations, in particular large organizations, function. Ideally, greater importance will be given to happy and healthy relationship networks which are acknowledged and fostered to the same degree as are organizational structures. I say this, not so much because it is 'a good thing for employees'—although undoubtedly it would be—but because there is clear evidence that healthy organizations with happy employees enhance profitability (Quick & Quick, 2004).

ORGANIZATIONAL FEAR AND A FROZEN SOCIETY

As business psychologists we, at Campbell Keegan, have conducted research studies with very diverse groups of people and within many different organizations since the early 1980s. The purpose of these studies is, in essence, to help organizations to understand their partners, consumers, the general public, their competition, particular sectors of society—depending on the purpose of the study. In turn, this understanding helps to shape the ways in which these organizations plan and move forward. In particular, we have specialised in the areas of wellbeing and finance and therefore we have followed the psychological impact of the recession with particular interest.

A number of themes have become apparent over the course of the last 4-5 years as the effects of the recession have deepened and spread throughout UK society. These themes inevitably reflect the mood of both individuals and organizations, in that individuals work within organizations and organizations relate to individuals in our connected environment.

Uncertainty and contagion

Research carried out by Campbell Keegan with the general public in late 2008/early 2009, during the early stages of the economic downturn, unearthed some interesting shifts in perspectives amongst the general public. New patterns of consumer thinking and behavior seemed to be emerging. Respondents talked about the inevitability of a restructured society in the wake of recession, of new priorities, moving away from excessive materialism, a return to community and family values. There was a sense that the downturn, frightening as it was, might prove to be a 'wake up' call to the world. It was apparent that respect and trust in financial institutions was probably lower than it had ever been. If these institutions were to regain consumer trust, they needed to start by understanding what consumers were feeling, wanting and demanding of them in the future (Campbell Keegan, 2008).

The research confirmed our hypotheses; that uncertainty dominated most people's thinking and that uncertainty is the stuff of nightmares for many of us. People described the 'caution', 'fear', 'paralysis, and 'unpredictability' of their current lives. According to neuroscientist, Johan Lehrer, we are much more comfortable with risk than we are with uncertainty (Lehrer, 2009: 196).

Molly, a 48-year old shop assistant voiced the fears of many others.

> *I am almost frozen about what to do. I'm so worried I might lose my job, have to re-mortgage, but I don't know what's right… I wish I could see into the future.*

Seeking meaning in familiarity and simplicity

Extreme feelings of uncertainty breed a flight to the familiar. There was a sense of seeking out safe, predictable situations and avoiding the unknown or risky. Long held friendships, the family unit, the home and habitual holiday destinations assumed exaggerated importance. Consumers claimed they were more likely to seek advice from family or friends and to avoid 'untrusted' financial experts.

The research also indicated a trend towards (or back to) traditional gender roles and traditional values, reflected in the popularity of 'Mad Men', 'Life on Mars' and 'The Boat that Rocked', which portrayed a sentimental view of a world of 'certainties' and simpler roles; workers vs. bosses, the establishment vs. the counterculture, men vs. women. Some men seemed to be embracing the 'new' gender stereotypes with some gusto.

> At the end of the day, it's the woman who has the children, so the man's job is more important. It does get back to the little wifey at home. (Paul, 26)

Perhaps more surprisingly, some younger women also seemed hungry to fill this traditional role, using language that would have seemed outmoded a decade ago, but which seemed to resonate with the prevailing economic climate.

> What I want is what my Mum has—the 4 by 4, the house and two children. I'm lazy… I'm prepared to admit it. I want marriage and a husband who looks after me. (Marie, 22)

As the recession bit deeper, there was more evidence that old fashioned sexism was reviving, along with a widening of the pay gap between men and women. A survey funded by the Chartered Management Institute (BBC, 31st August 2011) found that the income differential between female and male managers widened from 2010 to 2011.) At the same time, there appeared to be a growing pressure on women to be valued/to value themselves in terms of appearance, as evidenced by the inexorable rise in celebrity culture and cosmetic surgery which increased by 5.8% in 2011 (Guardian 30th January, 2012).

These societal changes in gender and power relations can alter the cultural dynamics, both inside and outside an organization and, in doing so, shape the organizational culture.

The cult of blame

In recessionary times, 'blame' has become something of a cultural football. Often it is at its most vitriolic when aimed at loosely defined 'others', e.g., 'the unemployed', 'asylum seekers' and 'single mothers'. It is not clear at whose door recessionary blame should be laid to rest, although bankers are most commonly blamed. The feeling of powerlessness and now anger at the 'greedy and incompetent financial institutions and a government that allowed it to happen' is proving a lethal cocktail. This notion of 'the Evil them' allows a comforting distance between 'the baddies' (the perpetrators) and ourselves (the innocent victims). It absolves us from responsibility. Self righteous anger and blame are natural psychological refuges at present but, paradoxically, they may be, in part, what keeps us, as a nation, from moving forward.

Doors of opportunity

In our research, many people described feeling physically and psychologically trapped by the recession. However, the research also highlighted a minority of, mainly young, people, both men and women, who saw opportunity in the doom and gloom (Campbell Keegan, 2008). For these individuals, the recession offered liberation. The rule book had changed and they no longer feel compelled to follow traditional paths. The current climate suggested radically new possibilities and/or could act as a catalyst for change. Moving job, shifting country, a major lifestyle shift or partnership change, and especially downshifting, all seemed possible.

Shifting perceptions and expectations

However, as the recession deepened, the mood of the country began to change. Research carried out in the last year (Campbell Keegan, 2011) suggested a shift in perspective, with many respondents, from different walks of life, expressing their inability to envision a positive future. They talked about feeling out of control of their lives, expressed uncertainty about their future and that of their children and many voiced a sense of passivity, even futility. Increasingly there seems to be an 'acceptance of the inevitable'; we will have to deal with whatever hand is dealt to us because control over their own lives is limited. There was also talk of 'getting back to normal', and by that respondents meant returning to pre-2008 days which were seen, retrospectively, as carefree and affluent. For many, this was ambitious enough. Changing society could wait.

But the cracks in society still show, even if we choose not to focus on them. In the immortal words of Leonard Cohen, *"There is a crack in everything. That's where the light gets in".* If we do not follow these cracks and learn from them, then we will simply repeat old patterns, perhaps with the same or worse consequences.

So far we have meandered between home life and organizational life because, as individuals, we necessarily move from one context to another and, to an extent, bring our hopes, fears and concerns with us. However, we will now focus more specifically on organizational life and the effects of recessionary pressures on the individuals who work within organizations, many of whom are struggling to retain optimism—and their jobs—in this time of uncertainty. In particular we explore how 'target culture' has impacted on working life and the sometimes counter-productive effects that have resulted.

PERFORMANCE TARGETS WITHIN COMPLEX ORGANIZATIONS

Our focus will be primarily on large organizations, the principles and structures that sustain them and the ways in which performance targets shape the interactions, both between employees themselves and between employees and the customers or clients they service. Of course not all organizations are the same; they differ by function, traditions, role, ethos and so on. However, senior managers who run large organizations tend to reflect contemporary management philosophies, to a greater or lesser degree. Fashions in management are just as prevalent as fashions in the high street. And just as some fashions flatter, others are disastrous.

I will argue that the ways in which many large private and public sector organizations operate needs re-examining. In particular, hierarchical, 'command and control' structures, which use poorly thought out and badly applied mechanistic targets, have fostered organizational cultures that are often toxic. I will examine the Mid Staffordshire NHS Trust scandal of 2010, in which it was estimated that more than 400 patients may have died through lack of care, whilst Trust managers attempted to adhere to their performance targets.

You may ask, "How is it possible that an institution that was set up for the clear purpose of helping the sick and dying could become so distorted in its practises?" At the very least, this example serves as a case-study on 'group think', a psychological term that refers to the process by which members within a group begin to form quick opinions that match the group consensus and over-ride their own views.

The extreme of this effect is mass hysteria (Janis, 1982: 2-9). To help prevent another such tragedy happening elsewhere, we need to try to understand how, and under what conditions, performance targets can create a contagious mind-set which works against both the aims of the organization and the good of the employees and/or their patients, customers or clients.

We are all familiar with target driven cultures. Some of us are subject to targets in our working lives. Others of us, as NHS patients, parents, drivers and so on, are subject to a wide range of targets, whether we are aware of them or not. Most targets are essentially quantitative measures. They can provide useful steers, especially when applied to simple goals, but they are limited in their usefulness. They are by nature reductionist, static and linear. Targets may remain constant even as organizational or client needs evolve—or even change rapidly, so that employees find it difficult to keep up. On their own, quantitative-based targets are inadequate for achieving a broad understanding of, and for attempting to work within, the complexities of dynamic, social organizational environments. This is not necessarily a problem, provided we recognise the limitations of the targets we set. Perhaps the most important limitation is that, if employees cannot meet the specified target, or if they do not support the targets because they believe them to be unattainable or irrelevant or counterproductive, then the target will fail to achieve its purpose—which is generally to measure and/or improve staff performance and organizational objectives.

Paradoxically, the targets themselves may be met, but they might not have achieved the goals they were designed to achieve. As Mark Scoular, Chief Inspector within the

London Metropolitan Police, explained at the height of the targets debate within the police force, targets may be met, but this does not necessarily mean less crime.

We no longer target criminality; we target how many stop-and-account forms we get during a shift…I fail to see how, with current key performance indicators, we are doing anything but fudging the real picture. (The Telegraph, July 8th, 2007)

So why are targets still ubiquitous throughout government and other large organizations, if they do not do the job they were designed to do?

A very eloquent explanation is offered by the theologian and philosopher, Alan Watts who, way back in 1954 wrote:

We think that making sense of out of life is impossible unless the flow of events can somehow be fitted into a frame-work of rigid forms. To be meaningful, life must be understandable in terms of fixed ideas and laws, and these in turn must correspond to unchanging and eternal realities behind the shifting scene…ideas and words are more or less fixed, whereas real things change….to define has come to mean almost the same thing as to understand…we resort to the convention of stills whenever we want to describe or think about any moving body, such as a train, stating that, at such-and-such times it is at such-and-such places. (Watts, 1954: 41-44)

Not much has changed since 1954! Quantitative targets appeal because, superficially, they appear straightforward, easy to apply and, on the surface, it seems easy to assess their effectiveness. They are comforting, promising certainty. Targets make a messy, contradictory world appear simpler, apparently more comprehensible. They allow us to reduce productivity to numbers—or so it seems. When targets 'fail', the response is often not to question the appropriateness of targets themselves, but to seek 'better', 'more precise' or 'more comprehensive' targets. Over the years, a great deal of effort has been expended on improving the efficacy of target systems, but with limited benefit (Boyne, 2002). Essentially this is because the 'perfect' target is an illusion and it is an illusion that has had some disastrous consequences for organizations, their employees and for the people they attempt to serve. Like all data, quantitative data needs analysis and interpretation. It is not an 'easy fix'.

The best way to illustrate this contradiction between the promise of 'quantitative certainty' and the reality is through examining some of the unintended consequences that can result from management performance targets, imposed in good faith, but 'adapted' to the situation on the ground.

Targets that hinder, targets that help

Targets, whether qualitative or quantitative, are not in principle 'bad', but they can become so if they are poorly designed, rigidly enforced and inadequately monitored. When meeting a target becomes prioritized over the task that it is designed to support, then there is a problem. In these instances, the target has become an end in itself. It becomes reified, fossilized. Such targets often promote a 'blame and reward' culture rather than a learning culture. In too many cases, this use of targets corrupts the organizational goals and purposes and can even undermine the whole ethos of the organization.

As social scientist, Alfie Kohn (1999: 48), points out in his punchy and provocative book 'Punished by Rewards', manipulating people through setting targets or by offering them incentives seems to work in the short term, but ultimately fails and even does lasting harm. According to Kohn, people actually do *inferior* work when they are enticed with money, grades or other incentives. As Kohn illustrates in his book, the more an organization relies on incentives, the worse it gets at achieving its purpose successfully and efficiently.

Psychologist and researcher, Joanna Chrzanowska, reinforces this point:

> *Targets are a current 'taken for granted', as many people work with an implicit behaviorist model in their minds about motivations. Ironically, the behaviorist model does work but largely in the negative—failing to meet targets is, in general, a de-motivator.* (Chrzanowska, 2010, personal communication)

Onora O'Neill's 2002 Reith Lecture (www.bbc.co.uk/radio4/reith2002), which addresses issues of trust, the effect of organizational targets and accountability, sums up this issue very succinctly:

> *I'd like to suggest that the revolution in accountability be judged by the standards that it proposes. If it is working we might expect to see indications—performance indicators!—that public trust is reviving. But we don't. In the very years in which the accountability revolution has*

made striking advances, in which increased demands for
control and performance, scrutiny and audit have been
imposed, and in which the performance of professionals
and institutions has been more and more controlled, we
find in fact growing reports of mistrust.

Clearly the problem does not lie with targets themselves, but in their application. Properly constructed targets, which are reviewed on an on-going basis and which are treated as diagnostic tools rather than ends in themselves, can make useful contributions to improving productivity, health and happiness within organizations. When targets are used in this way they act as signposts for the journey; they are not destinations. They can be flexible. If the particular targets are not working, that is, if they are not helping individuals and organizations to achieve their aims, then they can be modified—ideally by the people who are themselves subject to these targets. And just as targets can be flexible, they can also be qualitative. The word 'target' is often taken, by definition, to imply 'quantitative'; the *percentage* of crimes that have been successfully solved this quarter, the *number* of patients that have been seen in a hospital outpatient service within a specified time, *how many* cold calls an operative is expected to achieve each hour. It does not have to be like this. Professor Tim Blackman describes a constructive use of targets in which:

> *…targets are not regarded as reliable or valid ends in*
> *themselves but are re-framed as tracers picking out the*
> *key features of change as it happens. Employees become*
> *agents of change, alert to the feedback messages that*
> *these tracers send, modifying their own behavior with the*
> *understanding that outcomes are co-produced between*

themselves, colleagues and customers, with the resources each brings to the interaction. (www.radstats.org.uk/no079/blackman.htm)

This is quite a different definition of 'target' from that described previously. Unfortunately, the word 'target' has itself become contaminated through association with a variety of organizational disasters. Indeed the word initially conjures up an image of the firing line! Targets have become the 'bogey man' of management literature, often without a true understanding that it is the way in which targets are conceived, set up and used that is to blame, not the targets themselves. It is, after all, *people* who design and implement targets. The targets themselves are innocent! In order to avoid having to clarifying my use the term *target* throughout the rest of this essay, when I refer to targets I mean the rigid, quantitative, end-point style of target I referred to at the beginning of this section—unless I state otherwise.

As a broad generalization, simple targets are appropriate for simple objectives. However, few objectives in today's complex, multi-faceted organizations are simple. John Kay in his brilliant book, 'Obliquity' (Kay, 2010) sets out to convince us that complex goals are best achieved indirectly. He explains:

In general, oblique approaches recognise that complex objectives tend to be imprecisely defined and contain many elements that are not necessarily or obviously compatible with each other, and that we learn about the nature of the objectives and the means of achieving them during a process of experiment and discovery. Oblique approaches often step backwards to move forwards (p. 4)*… Problem solving is iterative and adaptive, rather than direct* (p. 9)*… High level objectives are typically loose and unquantifiable—though this does not mean it is not evident whether or not they are being achieved* (p. 41)*… The criteria of achievement are constantly redefined by great achievers* (p. 77) (Kay, 2010).

Kay goes on to give fascinating examples of how, in complex situations, oblique approaches can achieve greater success than direct approaches. I do not have the space to do justice to his arguments here. I recommend reading the book.

Targets in practice

Targets are not a recent phenomenon, although undoubtedly the prioritizing of targets has increased over recent years. I remember personally becoming aware of the effect of organizational targets in the 1980s. As business psychologists we were commissioned by a vehicle rescue organization to carry out a research project. The purpose of the project was to help steer an upcoming culture change programme focusing on customer needs. 'Putting Customers First' programmes were very fashionable at that time.

We initiated the research programme with a series of individual interviews amongst Members of the Board of this particular vehicle rescue organization. We explored the individual and shared objectives of the Board Members and probed the areas of conflict and convergence between them, particularly in terms of strategy development for the Customer Care programme.

Our aim was to help them to thrash out the bones of future organizational strategy. We then worked our way down—or up—the organization (depending on your perspective) eventually spending time shadowing service patrols who were sent out to rescue stranded motorists and, it was hoped, sort out their vehicle problems. We accompanied the patrols for several days, as the service engineers rescued stranded motorists from hard shoulders, towed them home, repaired vehicles, and such like. It was all good fun, as I remember. One of the great advantages of this type of ethnographic shadowing is that you are, on the one hand, a novelty and therefore it is usually fairly easy to get employees you are shadowing to engage with you and explain what is going on. On the other hand, you quickly become one of the crew, so that employees forget that you are *actually* there to observe and note their behavior. Both positions provide useful perspectives.

At that time, the vehicle recovery organization had implemented a customer pledge, backed by TV advertising. It guaranteed to send out a service patrol that would reach stranded drivers within the hour. This marketing goal was also linked to employee targets, which in turn fed into the staff bonus system. Simultaneously, competitor rescue organizations were introducing their own pledges; they promised waiting times that were less than an hour. Competition became fierce in terms of which organization was most effective in meeting their self-imposed targets.

Does it sound reasonable that recovery services should measure their performance in terms of the speed with which they arrive with the motorist? Probably it does. Isn't every motorist's wish, when they break down, that the rescue service arrives as quickly as possible? Well, that is not exactly true. Every motorist wants their car repaired and on the road again as quickly as possible. Quite a different thing. The target that senior management had set did not take account of wily human nature. Human beings are endlessly inventive—and sometimes Machiavellian. A target can become a barrier to be circumnavigated, especially when it is unattainable.

With the best will in the world, service patrols could not always get the appropriate rescue vehicle to the stranded motorist within an hour. For instance, if the vehicle needed to be taken to a garage for repair, then a tow truck was required. However, tow trucks were in limited supply; they were not always available in the area and patrols could not magic one up. The service patrols were mindful of their bonuses and the advertising claims of the company. This put pressure on them to achieve the unachievable. With a bit of lateral thinking, the solution proved to be surprisingly straightforward. If the appropriate rescue vehicle was not available, then an alternative vehicle, which *would* arrive within the hour, could be sent out to the motorist—*regardless of whether or not it had the equipment needed to deal with the problem.* So, a mechanic on a motorbike might be sent to a motorist who needed a tow truck. The target was met, but the hapless motorist might have to wait another hour or more before the necessary tow truck arrived. Ironically, this could mean an even longer wait for the motorist than it would have done if no 'fill in' vehicle had been sent. In this case, meeting a target which was designed to improve customer service actually had the effect of delivering poorer customer service. This example may seem amusing now, as we chuckle at the ingenuity of the service patrols—though it was probably not a great joke to the stranded motorist at the time—had she been aware of it.

Being stranded by the side of the road is inconvenient, but generally not life threatening. However, the effects of ill-conceived or un-monitored targets can be much more serious. As we touched on earlier, a scandal erupted within the Mid Staffordshire NHS Foundation Trust hospital, which would be difficult to believe if it had not been so thoroughly investigated after the event. The Times Newspaper summed up the situation.

> *Patients were routinely neglected or left 'sobbing and humiliated' by staff at (the) NHS Trust where at least 400 deaths have been linked to appalling care. An independent inquiry found that managers at Mid Staffordshire NHS Foundation Trust stopped providing safe care because they were pre-occupied with government targets and cutting costs.* (The Times, 25th February 2010)

In case the report above is mistaken for an instance of journalistic excess, it is worth including an excerpt from the Mid-Staffordshire NHS Foundation Trust Review of 29th April 2009, which investigated the case. Sir Ian Kennedy, Chair of the Healthcare Commission stated:

> *This is the story of appalling standards of care and chaotic systems for looking after patients.*

The executive summary in the government report which investigated Mid-Staffordshire NHS Trust continues:

A central theme of the failures at Mid Staffordshire Hospital Trust appears to be an over-reliance on process measures, targets and striving for Foundation Status at the expense of an overarching focus on providing quality services for patients.

How did this situation come about? The hospital staff were not evil people. They didn't set out to harm patients. I think it is reasonable to assume that they wanted to do their best for their patients. So how is it possible that targets could instil such a degree of group-think that no-one questioned—or even dared to question—what was happening? I cannot fully answer to that question. My supposition is that it was a mixture of fear and compliance with what had become cultural norms; that these factors induced a degree of myopia and mechanistic adoption of the rules. The culture was contagious; new staff were primed by the behavior of established staff. Cultural norms framed the way in which staff interacted and behaved. These norms were seriously detrimental to patients. Equally, targets certainly played their part. If you have any doubts that ill-considered and badly monitored targets can have destructive effects, you need look no further than this case. Mid-Staffordshire NHS Foundation Trust may be an extreme example, but sadly it is not an exception.

Mike Williams, an ex-CEO of an NHS Trust (now an academic), describes the difficulties of dealing with conflicting performance targets during a Norovirus epidemic at the NHS Trust hospital where he worked. Staff were required to juggle waiting time targets and simultaneously manage targets for emergency admissions. However they could not reduce patient demand by cancelling elective patients for fear of breaching other targets and they were not permitted to close the hospital although staff were also catching the virus, which put further pressure on resources. The fear of breaching targets, even in an emergency situation, meant that common sense could not prevail (Williams, 2010: 605-613).

You might say, '*Ah well, the targets were wrongly set in the first place. If the targets had been patient- centric, if the goal had been 'better care for patients', then these awful consequences would never have happened*'. There is some truth in this, but it is a dangerous path to follow because the underlying assumption is that, "*if only we can find the perfect target and we can stick with it, all will be fine*". This assumes a static, predictable world. Few, if any, of us work in that type of environment, nowadays if, indeed, we ever did.

Situations change. Life is messy and complex, particularly in fast response environments such as hospitals, and it is likely to become more rather than less so, as expectations of response times are accelerated through modern communications. A great many decisions in life are, essentially, "qualitative" that is ,they are complicated, fluid, difficult to pin down, they change over time, they depend on context and so on. Increasingly we are required to react, communicate, and think more quickly and in an improvisational manner within organizations as well as outside them. Targets can only ever be steers, useful for pointing us broadly in the right direction. If we believe that targets on their own hold the answer to efficient and effective organizations, then we are on course for disaster.

PROBLEM SOLVING WITHIN THE POLICE SERVICE

In 2010 we were commissioned to carry out a study aimed at exploring how problem solving within the UK Police Service might be improved. The joint commissioners were the Royal Society for encouragement of the Arts, Manufacturing and Commerce (the RSA) and the National Policing Improvement Agency (the NPIA). The exploration was conducted from the perspective of organizational development. This was quite a daunting brief. We had not worked with the Police Service before and there was something of the 'teaching your grandmother to suck eggs' about the project. In addition, there was a limited budget and we were attempting to serve two different masters—the RSA and the NPIA—each with quite different perspectives, styles and agendas.

To give a brief background to the study: Systematic approaches to problem solving have been used by the UK Police Service for decades and there is clear evidence that they increase the likelihood of reducing crime and disorder. For instance, SARA (Scanning, Analysis, Response and Assessment) has been used by the Police Service for many years. In essence it prescribes a simple, staged process of dealing with a problem. However there are many complex versions of SARA and in practice it can be difficult to implement because the situations Police Officers find themselves in are often volatile, unpredictable and unstructured.

However, it was apparent that, on many occasions when SARA, or a similar approach, could have been useful, it was not used. The key question, therefore, was why these problem solving frameworks were not used more frequently when it was clear that they could be effective? What were the cultural and organizational factors that discouraged Police Officers from employing them—and could this be changed?

We conducted intensive research amongst Police Officers and the general public in three diverse geographical areas within the UK, employing a mixture of extended face to face interviews, group discussions, shadowing and interacting with Police Officers and the general public (several times on drug raids as they broke down doors and once when we helped to carry a hefty sniffer dog over broken glass so he could do his job without getting glass in his paws!). We attended police briefings, accompanied Police Officers on arrests and when interviewing suspects. We talked in depth with Police Officers at different levels of seniority, from Chief Inspectors to PCSOs. We also conferred with related agencies such as the Fire Service, the NHS, housing associations and probation services who worked in partnership with the police.

This was one of the most interesting projects we have undertaken and we gathered a huge amount of data. Along with two other research teams, who were working independently but in parallel with us, we presented our findings and recommendations to around a hundred and fifty Police Officers and academics in a day long Symposium at the RSA.

For the purposes of this essay, I will focus solely on performance targets and their role within police work. Our study was exploratory. However, it was sufficiently extensive to highlight some of the ways in which targets can pervert their purpose; how they can sometime achieve the opposite of what was intended. This perversion has similarities to the way in which targets can be seen to operate in other organizations even where, as in the case of the vehicle rescue organization, the nature of the business is quite different.

Problem solving was our focus, so were looking at targets within this context. Very quickly it became clear that 'problem solving' meant different things to different Police Officers and/or in different situations. Broadly there were three different interpretations of 'Problem Solving':

1. **Problem Solving is what the police do**. From this perspective, all of the activities that Police Officers are involved in can be considered to be problem solving. This was the loosest—and therefore the least useful—definition, because it encompassed every aspect of police work and there was no easy way of differentiating between *good* and *poor* problem solving.

2. **Problem solving is an approach aimed at achieving pre-defined police targets**. Since 5th March, 2009, the only official national police target set by the UK government is to increase public confidence by 15 percentage points (http://www.community-safety.info/56.html). In spite of this, the Police Service is still driven by targets. There are many reasons for this, including the need for senior officers to prove their effectiveness and the fact that officers themselves are evaluated in terms of meeting targets, rather than successfully solving problems.

3. **Creative problem solving.** A creative and improvisational approach to solving problems is directed towards understanding the problem. It uses past knowledge and experience to make appropriate judgements and to decide on the best way of addressing the underlying issues—as well as the problem symptoms. This approach often involves:

    i.   Broad analyses of the problem *within its context;*
    ii. Identifying the *root causes* as well as the symptoms;
    iii. Ideally *changing the conditions* that prompt recurring crime;
    iv. *Multi-agency activity* (probation services, fire services, local council etc.)—where appropriate;
    v.   Looking at the problem from *different perspectives;*
    vi. Assessing the effectiveness of the problem solving approach to enable *iterative learning*, shared with colleagues;
    vii. Persistence over the *long term.*

    The SARA model (or more sophisticated versions of SARA), broadly mirrors the creative problem solving route, outlined above.

Clearly, some problems that Police Officers have to deal with are simple and others are complex. The police approach needs to be adapted accordingly. However, it is clear that approach (i) is too unstructured and hit and miss to be more than randomly successful whereas approach (ii) may be appropriate when the problem to be tackled is straightforward, easy to define and easy to measure in terms of the level of success of the outcome. For instance, success in terms of directing traffic away from a burst water main or reducing the level of littering or dog fouling in the street, are relatively simple problems to define and the level of success in achieving these goals can be monitored reasonably accurately.

Approach (iii), however, would seem to be the most relevant approach when problems are complex, multi-faceted, when they are difficult to define and/or they are developing over time. The way in which these factors compound one another is much discussed amongst Police Officers, so the question we asked ourselves was, "Given that there is widespread familiarity with the SARA approach and its efficacy, why do Police Officers not automatically adopt a SARA approach when the conditions above suggest it?"

A number of reasons for this were identified in the research. The two that I will focus on here are firstly, the predominance of 'Fast Response' in the Police Service and secondly, the effects of performance targets on police behavior.

Fast Response culture

The British Police Service is built around quick response. Clearly this is a necessity and an organizational priority for the police. Many officers are attracted to the Service because they like this style of working—and police training amplifies this tendency, especially within Response Teams that are charged with responding quickly in emergencies.

However, 'Fast Response' can, in turn, encourage a 'quick fix' mentality, for example by focusing on and dealing with symptoms rather than underlying causes. This approach can work against effective, long-term problem solving because, too often, unless the root cause is tackled, the problem will re-occur. Teams that are target driven may focus on *doing and being seen to do,* rather than achieving a long term solution. In addition, and most importantly, a more painstaking, reflective, analytical approach is often not rewarded within the Police Service. *Status and recognition often come from meeting targets, not from solving problems.*

This is a hard nut to crack, particularly amongst police teams that are trained to react quickly and effectively in emergencies. The challenge is to ensure that Police Officers can quickly assess the needs of a particular situation, react quickly when the situation demands it, but act in a more reflective fashion when that is more appropriate. This sounds straightforward, but of course the reality is that officers sometimes have to make split-second decisions, with limited information, in very tense and volatile situations.

At Campbell Keegan, we have worked with a range of organizations that employ performance targets. In all cases, the targets have affected staff performance. However, at the stage where the targets are implemented, it is very difficult to predict exactly what these effects will be. Within the police study, three significant effects of performance targets were evident. Each of these had an impact on Police Officer behavior.

Currently, if a crime is not defined as a *priority* within a particular Police Force and/or if it is not measured by a target, then it is less likely to be detected and less likely to be addressed. In this sense, "*A crime without a target may not exist.*"

A poignant illustration of this effect was the way in which drugs were viewed in one geographical area of the country in which we conducted research. It was clear that drugs were a problem in this area; we had accompanied drug squads on raids, in which they busted and arrested drug dealers. Nonetheless, the mantra within the particular Basic Command Unit (BCU) was "There are no drugs in xxx". I kept hearing this statement—followed by a laugh—and eventually I cornered one of the officers whom I had been shadowing. Reluctantly, he explained. "We have no drugs in xxx. If we had drugs, we'd need a drug squad, but we can't afford a drug squad. And if we did have a drug squad, we'd have targets, but we wouldn't be able to meet the targets because there is no way of controlling the drugs. Drugs are at the root of most crime here, but there is nothing we can do about it." The best solution therefore, was to deny the existence of drugs. No drugs. No targets. No problems.

In this way performance targets can both influence the definition of crime and also determine whether or not that crime is considered worthy of prioritizing. Targets are insidious. They occupy no moral high ground. They can work just as effectively against crime reduction as for it.

A Police Officer called to a 'domestic dispute' generally has two choices. He/she can arrest the suspect, charge him or her, and take them into custody, even if this is against the wishes of their partner (who may have initially called the police). Arresting the suspect probably means he/she will spend a night in jail, will be absent from work, possibly get a criminal record and there may be other repercussions. In this instance, the police officer will have followed the rule book, met a target, but will probably not have exercised much professional judgement.

Alternatively, the officer could talk to the offender and the complainant, consider the context in which the incident occurred, evaluate the severity of the incident, establish if it was a first call out, seek assurances of no repeat incident—and then make an informed judgement about whether or not to charge the offender on this occasion. However, going down this route makes the police officer vulnerable. There is a victim, there is a crime and there is a suspect, but there is no tangible result. Therefore the officer must log the incident as an 'unsolved crime' and this will not look good on his record—or on the BCU targets. In this instance, performance targets can act as a disincentive for the officer to use his judgement and to take responsibility for solving what might be a relatively minor problem.

Emphasizing *rules*, rather than *principles*, encourages a mechanistic response. Officers cannot use the experience, skills, intuition and knowledge they have honed over years of dealing with the general public. Given that most of our judgements involve emotional, bodily, as well as intellectual, input aren't we missing a trick here? As John Kay (2010: 168) puts it, "By downplaying genuine practical knowledge and skill in pursuit of a mistaken notion of rationality we have in practice produced wide irrationality—and many bad decisions."

Meeting the target may not solve the problem

A group of teenagers hangs out in a patch of wasteland near a high wall. On the other side of the wall is a showroom for a Mercedes Benz dealership. A collection of expensive cars is regularly parked in the forecourt of the dealership. At some time during the evening, a group of drunken teenagers start throwing bricks over the wall. The bricks hit the cars and cause damage. This happened irregularly, roughly once a month. Each time the police are called by local residents. Sometimes the officers catch one or two of the young people, but this does not deter the others for long. After a while, the trouble starts again. This goes on for some months. Both neighbors and Police Officers are frustrated, as indeed is the dealership. Then a Senior Officer decides to investigate the site. He walks around the wasteland and the forecourt of the dealership and then asks, "Why are all these discarded bricks lying around?"

No one knows. Junior officers are tasked with clearing away the bricks. Miraculously, over the course of the next few weeks, there are no attacks on the cars. The Senior Officer had asked the obvious question; one that no-one else had asked. He had realized that the attacks weren't premeditated but were the impetuous actions of bored, rebellious teenagers. Remove the bricks and you remove the temptation. Of course it is probable that the teenagers would find another site, or do something equally anti-social, but it did break that particular cycle of repetitive criminal behavior. Quick fix responses (police repeatedly being called out and arresting/cautioning a few teenagers) which contributed to the police targets but did not provide a long term solution, had been converted into creative problem solving—which *did* solve the anti-social behavior—at least locally.

In summary

Targets serve purposes. Applied to simple goals, they may be *good enough* to assess the effectiveness of achieving those goals or to gain some measure of productivity. Managers traditionally use targets to measures sales, bonuses, and so on, or to justify their role, their behavior and their success to their senior managers or to the general public. Targets offer an illusion of control. But more often than not targets (at least as the sole method of evaluating) do not serve the purpose they ostensibly set out to serve, especially when we are dealing with complex situations, where outcomes are unpredictable and evolving. Targets are too static, too unforgiving. They are unable to take account of the iterative processes of exploration, reflection and experimentation which are necessary when steering complex decision making. Essentially targets, on their own, are too crude and reductionist to help us to make sense of multi-layered problems in our complex world. So what to do? Do we abandon targets? Or we look for alternative or complementary approaches?

THE INSIDIOUS QUALITATIVE-QUANTITATIVE DIVIDE

Before exploring alternatives—or complements—to performance targets, I want to shine some light on the whole notion of qualitative and quantitative ways of understanding the world. We tend to view the division between qualitative and quantitative spheres as self-evident, a fact of life, laid down by the hand of God or the rule of nature. Rarely do we question it. We take it for granted that 'quantitative' is essentially to do with 'hard data'; measurement, percentages, proportions, whereas 'qualitative' data is concerned with 'soft data'; how, why, patterns and connections. Performance targets are therefore assumed to fit more neatly within the quantitative camp.

In many business environments, qualitative and quantitative data are treated as if they are fixed entities set up (by whom?) in opposition to one another, rather than being useful tools for understanding the world. The Western mind-set is predicated on the belief that every object, thought, feeling, is defined in relation to what it is not. We have mind vs. body, classical science vs. quantum, mechanical vs. organic, me vs. the world. Thankfully these sharp divisions are beginning to blur as the complexity sciences and quantum physics continue to challenge our preconceptions about the rigidity of objects, thoughts and feelings. However, we still use opposition as a way of defining—and indeed understanding—our world. So it seems natural for us to treat the quantitative-qualitative divide *as if* it is reality.

The social constructionist, John Shotter (1993: 11), coined the phrase 'rationally invisible' to describe phenomena that are so self-evident, so taken for granted, that we do not even 'see' them. The supposed divide between qualitative and quantitative ways of understanding the world seems to be one such 'rationally invisible' phenomenon. Because we regard the division as given, immutable, it is too easy to position ourselves on one side of the fence or the other.

Either we view qualitative data as rich, connected, 'alive' and dismiss quantitative data as superficial and soul-less.

Or we view quantitative data as accurate, precise, factual, whereas qualitative data is woolly and subjective.

I am exaggerating. However, it is easy to forget that there is no insuperable barrier between qualitative and quantitative understanding; they are just constructs that we impose on 'reality', different lenses through which to view the world. And, because the division is assumed—'a fact of life'—it is a short leap to perceiving 'quantitative' as *in opposition* to 'qualitative' and vice versa. And so, the familiar battle lines are drawn.

The qualitative-quantitative divide grew largely out of post-industrial Western culture. It is a useful tool, a metaphor. The Milky Way (probably) does not change according to whether we count the stars or write a poem about it—it is our way of perceiving the Milky Way that changes. In the same vein, qualitative and quantitative are not intrinsically different. We impose the distinction between them, rather than it being a function of the natural order.

There are many advantages in bringing quantitative and qualitative mind-sets together. For a start, we can build up a more holistic and multi-faceted picture and, at best, appreciate the detail as well as the 'big picture'. How might we achieve this? Can we incorporate qualitative and quantitative perspectives within performance assessments, so that we can capitalise on these broader perspectives. Perhaps we need to look further afield?

Alternative perspectives on the world

Not all cultures think in the either-or, qual-quant way that we have become accustomed to. Australian aborigines have a sophisticated cultural life. Religion, history, law and art are integrated in complex ceremonies which depict the activities of the ancestral beings that created the landscape and its people. There are prescribed codes of behavior and responsibilities for looking after the land and all living things. Aborigines view themselves as part of the land, not in opposition to it. Similarly, many Eastern cultures emphasise the *connections* between people, God, the land, culture, rather than the divisions (Watts, 1969). Where there is a perception of difference, the parts are often seen as necessary elements of a whole, for example Yin-Yang which describe how polar opposites or seemingly contrary forces are interconnected and interdependent in the natural world.

Many decades ago, when I was studying undergraduate psychology, I became interested in 'Eastern thinking' (as it was then called); Buddhism, Hinduism, Sufism. I came across Alan Watts, a writer and philosopher who translated Zen Buddhism into a form which made it more accessible to a Western audience. "The Book on the Taboo against knowing who you are", was first published in 1969, but it still seems extraordinarily prescient today. Watts draws on both Newtonian science and Zen Buddhism to challenge prevailing Western perceptions of reality.

When I first encountered Watts's writing, I was bowled over by the notion of viewing the world as fluid, indivisible, a pattern of movement, which we, as 'individuals', come *out* of—in much the same way as the wave *comes out* of the ocean and sinks back into it—because we are a transient expression of the whole realm of nature. Watts introduced me to the idea that everything in the world is connected. It was the first time I had come across these ideas; they were rather eccentric in those positivist days in which the emphasis was on humans as superior to nature; conquerors of the universe. Nowadays, the concept of a connected world—Gaia—is mainstream, and we are forced to appreciate how climate, consumption and communication feed off one another.

Watts's ideas seem to me to have particular relevance to the qual-quant division—as well as being entertaining—and so I have quoted him at length below:

I have sometimes thought that all philosophical disputes could be reduced to an argument between the partisans of "prickles" and the partisans of "goo". The prickly people are tough-minded, rigorous and precise, and like to stress differences and divisions between things. They prefer particles to waves, and discontinuity to continuity. The gooey people are tender-minded romanticists who love wide generalizations and grand syntheses. They stress the underlying unities, and are inclined to pantheism and mysticism. Waves suit them much better than particles as the ultimate constituents of matter and discontinuities jar their teeth like a compressed-air drill. Prickly philosophers consider the gooey ones rather disgusting—undisciplined, vague dreamers who slide over hard facts like an intellectual slime which threatens to engulf the whole universe in an undifferentiated aesthetic continuum. But gooey philosophers think of their prickly colleagues as animated skeletons that rattle and click without any flesh or vital juices, as dry and desiccated mechanisms bereft of all inner feelings. Either party would be lost without the other, because there would be nothing to argue about, no one would know what his position was, and the whole course of philosophy would come to an end…

Historically, this is probably the extreme point of that swing of the intellectual pendulum which brought into fashion the Fully Automatic Model of the universe, of the age of analysis and specialization when we lost our vision of the universe in the overwhelming complexity of its detail. But by a process which C.G. Jung called "enantiodromia", the attainment of any extreme position is the point where it begins to turn into its own opposite—a process that can be dreary and repetitive without the realization that the opposite extremes are polar, and that poles need each other. There are no prickles without goo and no goo without prickles. (Watts, 1969)

What Watts is saying about the differences between *prickles* and *goo* can easily be transposed into the way in which qualitative and quantitative mind-sets operate within different organizations or different facets of organizational culture, as follows:

Quantitative mind-set (Prickles)

- An individual slots into the organization and understands their allocated role;
- Objectivity, dispassion and distancing are privileged;
- The focus is on 'things'; roles, hierarchies, targets etc.;
- Constancy and continuity are prioritized;
- Linear, mechanistic planning, measurement and control are favoured;
- Structures are created, and then their creators distance themselves from the structure, i.e., the organization is perceived as having a life of its own;
- People often feel powerless in the face of these structures.

Qualitative mind-set (Goo)

- Working styles can be fluid, organic, in the moment;
- An individual builds diverse relationships across the organization, depending on work needs;
- There is a focus on relationship *between* things;
- Emotion and feeling are core values;
- Change and new thinking are encouraged;
- People feel they have some control and power over their working lives; that they are part of the whole;
- Organizational style is epitomized by improvisation, complexity, autonomy, egalitarianism.

Of course this division is ridiculously over-simplified, but I want to highlight the distinctions. There is clearly a difference between prioritizing structure and prioritizing relationships. Some organizations and/or individuals tend to put more emphasis on structure, others emphasise relationships. For instance, senior managers, tasked with making people redundant, may retreat into structure, viewing those who are to be made redundant as 'cogs in the machine'. They may find it easier to deal with employees in this way rather than addressing the messy human stuff. Other managers may focus on relationship; trying their best to help those whom they are obliged to make redundant, because they feel it is part of their job to do so. Crudely we could say that the first manager is adopting a more disengaged or quantitative approach and the second one is adopting a more connected or qualitative approach. Organizations as a whole may have a particular style and within different organizations different styles of interaction are encouraged. The current vogue is to veer towards a quantitative 'norm'.

This rather caricatured division between quantitative and qualitative aspects of organizational life is important, because it underpins target culture. If we start with the assumption that there *is* actually a split between the two ways of seeing the world (rather than it just being a useful model)—and then we prioritize one side of the split over the other, we create imbalance. Prioritizing quantitative perspectives on the world and suppressing qualitative perspectives has led directly to the dominance of target cultures that we have experienced in the last decade or so.

Perhaps the healthier option is on-going tension between the two; qualitative and quantitative, discouraging the 'stuck-ness' that results from comfortable, self-reinforcing beliefs.

The most important message I take out from Watts's writing above, is the assertion that prickles and goo (or quant and qual) are interdependent—"*there are no prickles without goo and no goo without prickles*". This is something that we, in our either-or world, find difficult to retain in our minds: We are more likely to see prickles and goo as at odds with each other, eternally at war. But it is a war that neither can ever win.

In a time of stress and lack of clarity, such as we are currently experiencing, we are almost hard-wired to seek certainty—and quantitative measures provide an illusion of certainty. This can have unintended and sometimes destructive consequences, particularly within large, complex organizations. The attempt to 'lock down'; to impose strict targets, to monitor and quantify employee behavior is often counter-productive because it prevents people from using acquired knowledge, skills, improvisation and intuition in dealing with the unexpected—and in large, complex organizations, the unexpected is often the norm.

THINKING QUALITATIVELY WITHIN ORGANIZATIONS

The earlier part of this essay focused on the potential pitfalls of applying simplistic, ill thought out and poorly monitored quantitative targets within complex, evolving organizations. By complex organizations, I am talking primarily about large, multi-functional organizations with many employees, often located in different sites or departments, with discrete processes, rules, strategies and structures. Within these organizations, problems and issues are typically emergent and often unpredictable. The case study exploring problem solving within the Police Service, described earlier, is a good example. Police Officers may be working within extreme situations in which they have to make quick decisions with limited information and a poor understanding of the context in which they are operating. Even in less extreme situations, working effectively within complex organizations calls for finely honed human judgement based on past experience and an assessment of current needs—which in themselves are often changing. Employees need to address the immediate task at hand but, at the same time, understand this task within the context of wider organizational aims and objectives. This can be quite a challenge.

Qualitative approaches can provide a different perspective within organizations, especially when integrated with quantitative approaches and 'owned' by employees themselves. They can help us to steer and improve organizational productivity whilst, at the same time, acknowledging that much organizational behavior is, of necessity, improvisational. However, this is not an easy option. It needs on-going attention and requires skilled input to set up, steer and interpret. On the other hand, qualitative approaches enable us to work more effectively with the dynamic nature of complex organizations, to explore how the organization is developing, how different influences compound one another, how relationships between different work groups or levels in the organization or departments are changing and how this impacts on staff performance. This understanding can provide fodder for future organizational development— much of which may be driven by employees themselves.

Greater emphasis on qualitative methods of steering and evaluating productivity within complex organizations is likely to improve working practices and performance—for the simple reason that qualitative approaches mirror the way in which complex organizations function, i.e., they are non-linear, , emergent and encourage connection between different employees, levels and departments. Qualitative methods can be adapted to the changing situation; they involve human judgement and, at best, they examine both the big picture and the detail. The emphasis is on employees working together. Just as we would be hard pushed to measure personal happiness by means of a quantitative questionnaire (although many researchers would disagree with me on this), gauging 'the health of an organization' is, I believe, difficult to achieve through quantitative measures alone.

The concept of *Qualitative Mind*

So far, I have used the terms 'qualitative methods' and 'qualitative thinking' quite loosely. But what do we mean by 'qualitative thinking' and how might it help us in organizational development?

The psychologist and philosopher John Dewey introduced the phrase *Qualitative Mind* to describe "the interaction and exchange between the dynamic qualities of a live person and the equally dynamic qualities of the person's experiential world" (Brigham, 1989). More recently the term has been used by the author and qualitative researcher, Joanna Chrzanowska to describe the processes of qualitative market research (qualitativemind.com). Broadly, *Qualitative Mind* describes the human ability to generate knowledge (intellectually, bodily, emotionally) as we selectively incorporate information and integrate it with our existing knowledge, through a process of selection, reflection, monitoring, feeling, intuition and evaluation as an iterative process—both individually and with others. *Qualitative Mind is an on-going process, not a 'thing'.* This may sound complex. However, in practice, it is what we all do every waking moment—we just take this very sophisticated way of making sense of the world for granted (Keegan, 2011: 67-80) and in doing so these processes have become 'rationally invisible' (Shotter, 1993: 60).

I have been a qualitative researcher and organizational change consultant for more than thirty years. In that time I have come to regard *Qualitative Mind* as a way of *being* as much as a way of working. This particular way in which we make sense of the world and interact with others can be summarized below as:

- Being part of a process; a participant and observer at the same time;

- Helping to understand the nature of the problem;
- Accepting that knowledge is constantly evolving;
- A state of curiosity, openness and engagement;
- Using all our faculties; intellect, feeling, beliefs, intuition, experience;
- Encouraging creativity and diversity;
- Looking for patterns and connections;
- Iterative learning;
- Improvisational; being able to incorporate new information, to adapt and change;
- Reflexivity, rigour and discipline.

The psychologist, Professor Judi Marshall describes this state of mind succinctly as "Living life as inquiry" (Marshall, 1999).

This description of *Qualitative Mind* is doubtless incomplete and you may question the way in which I have attempted to define it, but it is probably good enough for our purposes. In my view, the qualities outlined above more or less encapsulate the way in which we engaged in qualitative inquiry or qualitative thinking at its best. These attributes and ways of making sense of the world have been honed over the decades by qualitative research practitioners. Of course researchers do not have a monopoly on these qualities. However they have been practiced and developed by qualitative researchers with discipline and rigour and, I believe, they are skills that need to be employed more widely, particularly within organizational contexts.

The classical research model views research as a linear, staged process—the product of *Rational Mind*. Data are gathered like 'things' or 'findings', then sifted, categorized and sorted before being presented to the client in a structured and logical way as 'Conclusions and Recommendations'. This model of Mind was built on the assumption that all thinking is conscious, rational and sequential. However, contemporary neuroscience is challenging this model; it has become increasingly apparent that this is not the way our minds work—or at least it is not the *only* way our minds work. In many ways, *Qualitative Mind* seems to be closer to the reality of the way we think than is *Rational Mind*.

Qualitative Mind is how people think, feel and act in *practice*, both individually and in groups. Human beings do not have separate compartments for emotion, feeling and thinking. These human qualities are all muddled up together and we cannot easily disengage them. Daniel Kahneman provides excellent examples of the distortions, delusions and deceits of the human mind in his wonderful book, 'Thinking Fast and Slow' (Kahneman, 2011). Our understanding of the people with whom we are working (colleagues, clients, employees, customers, stakeholders etc.) needs to acknowledge the messiness, contradiction, illogicality and creativity that make us human. And, at the same time, if we are in the role of researcher or organizational change consultant, we need to be able to make sense of this chaos and communicate with our colleagues and clients in ways that will help them to address the relevant issues. This is a hugely skilled task. It requires that we integrate qualitative and quantitative perspectives, the rational and non-rational, past experience, the current context, anticipation of the future, intuition, as well as feeling and bodily sensations. In short we need to be to be fully present as human beings.

Perhaps the most important aspect of *Qualitative Mind* is the capacity for emergent thinking; the process by which the brain constantly updates itself, making new meaning and confirming previous meaning, second by second, *in the present,* using all the faculties mentioned above. Ralph Stacey, professor at the Complexity Research Group at Hertfordshire University describes this as follows:

> *The process perspective takes a prospective view in which the future is being perpetually created in the living present on the basis of present reconstructions of the past. In the living present, expectations of the future greatly influence present reconstructions of the past, whilst those reconstructions are affecting expectations. Time in the present, therefore, has a circular structure. It is this circular interaction between future and past in the present that is perpetually creating the future as both continuity and potential transformation at the same time.* (Stacey, 2003: 10)

Our brains are structured in such a way that we cannot just absorb data without influencing its content. The brain automatically makes meaning. That's its job. In practice, as thoughts, feelings, hypotheses spring into our minds, we may backtrack to re-evaluate and shift our previous thinking, move forward, then back and so on. The emergent processes of *Qualitative Mind* can be pictured as more of a spiral or a series of iterative loops, like a spring, rather than a series of clearly defined staging posts. Of course, this process does not happen in isolation but is influencing and being influenced by others around us. This is a very different model than that of *Rational Mind.*

When we think about it, the skills discussed above are exactly the skills that are needed within modern day organizations. In a world in which uncertainty is a constant, we need people who have considerable knowledge and skills *but who can also improvise effectively and appropriately*. We tend to think of improvisation as a weakness, a lack of planning. It can be exactly the opposite in a situation in which planning is of limited value because the future is unknowable. Acting *appropriately* in the moment, being able to take on board the importance of history, context, the current demands and future expectations—with accuracy and confidence—is an essential skill—*and a definition of effective leadership.*

Look at the recent political scenario in the UK: Extraordinary events unfolded. Against all expectations, the Liberal Democrats—a third runner in the political race—shot to lead position on the back of a live television debate (according to some polls). All the parties involved, including the Liberal Democrats, were shocked. Just weeks before the election, after the plans, the speeches and the rhetoric had been prepared; they all had to be torn up. With the introduction of televised political debates between the different party leaders, politics in the UK will ever be the same again. The landscape has changed forever and the criteria for choosing political candidates have subtly shifted. The parties were forced to think *on the hoof*, to improvise, to constantly re-position themselves in relation to the other parties. This is not necessarily a long term benefit for politics and the British public (some would say it is mockery of democracy and a nod to celebrity culture) but in a neck and neck race, I would predict that the party that can improvise most convincingly, that can learn and adapt as they go along, will be the one to win the next election. This is *Qualitative Mind* at work.

ORGANIZATIONS AND PEOPLE

I n recent years, partly in reaction to the limitations
of target culture, much of the thinking and research
within organizational change has converged, to greater
or lesser degrees, on a rather broad common territory
which could loosely be called *psychological* or *relational,*
in that it prioritizes human aspects of organizational life,
rather than organizational structures and goals. This does
not mean, of course, that structure can be ignored, but it
does mean that human values are regarded as at least of
equal importance. Management theories and practices
that fall within this category include those from within the
following stables: complexity thinking and emergence,
behavioral economics, the 'learning organization', 'bottom
up' organizations and what I have called *Qualitative Mind*—
to name but a few. Some theories prioritize a way of
thinking to achieve change. Others have linked change to
a way of working. Both are important—and arguably they
are the same thing, in that thinking is a form of action. I
have briefly included some of these schools of thinking
below, in order to help fill in this relational territory.

Over the years we have been bombarded with the 'latest
and best' management fashions, which have come
and gone. We each need to develop the confidence
to use the approaches that are most appropriate to
the issues at hand and that utilise our own skills and
interests. For this reason I have included a taster of a
number of contemporary perspectives. I do not believe
that these approaches are mutually exclusive. However,
my particular interest lies within the field of complex
responsive processes of relating, which I believe more

accurately reflects lived experience within organizations, as opposed to reified models of organizational life.

The learning organization

The qualities that I sketched out in relation to *Qualitative Mind* overlap to some extent with the work that organizational guru, Peter Senge spelt out more than twenty years ago in his book *The Fifth Discipline* (Senge, 1990). Sadly, Senge's approach, whilst very compelling at the time, seem to have faded with the rise of target culture. He described the Learning Organization as:

> *…organizations where people continually expand their capacity to create the results they truly desire, where new and expansive patterns of thinking are nurtured, where collective aspirations are set free, and where people are continually learning to see the whole together.* (Senge, 1990: 3)

Senge, who described himself as an idealistic pragmatist, focused on introducing 'Systems theory' into organizations and bringing human values into the workplace. In our hard-nosed age, it is easy to dismiss these sentiments as over-idealistic. However, we have only to glance at the state of some large organizations today to understand the need to explore alternatives to target driven and high anxiety cultures. Often, it is not only misaligned targets that drive organizations but also 'group-think', discussed earlier, a mentality that can be observed within some sectors of the banking industry—and which makes it difficult for them to understand public fury over their behavior.

The dimension that, according to Senge, distinguishes the Learning Organization from more traditional organizations is the mastery of certain basic disciplines. Senge defines five dimensions that together contribute towards the development of the Learning Organization. These are:

Systems thinking: This is Senge's conceptual cornerstone. He stresses the importance of understanding and addressing the whole organization and the interrelationship between its parts. Senge—and more recently, and very eloquently John Kay (2010)—talks about how managers frequently apply simplistic frameworks to complex systems. We tend to focus on the parts rather than seeing the whole, and fail to see the organization as a dynamic process. A better appreciation of systems, Senge argued, will lead to more appropriate action. Some management thinkers, such as Ralph Stacey, would refute Senge's assumptions, arguing that systems thinking implies a bounded world whereas, in Stacey's view, emergence transcends boundaries (Stacey, 1996).

Personal Mastery: "Organizations learn only though individuals who learn. Individual learning does not guarantee organizational learning, but without it no organizational learning occurs". (Senge, 1990: 139). Senge viewed personal mastery as the discipline of continually clarifying and deepening our personal vision, of focusing our energies, of developing patience, and of seeing reality 'objectively'. This, he believed, went beyond competence and skills. He saw it as involving spiritual growth, in which people lived in 'a continual learning mode', but never 'arrived' (Senge, 1990: 142). This concept of personal mastery is akin to much psychological thinking, such as that of the psychotherapist Carl Rogers (1961), whose thinking shaped much early qualitative research work. I would argue that much of qualitative research and analysis involves the same processes, which I term 'emergent inquiry' (Keegan, 2009a, 2011).

Mental Models: These are 'deeply ingrained assumptions, generalizations, or even pictures and images that influence how we understand the world and how we take action' (Senge, 1990: 8). Philosopher, Donald Schon (1982) calls them a professional's 'repertoire'. John Shotter describes them as rationally invisible (Shotter, 2003). In order to change our behaviors or attitudes we often need to challenge our mental models; to 'turn the mirror inwards' and scrutinise our view of the world whilst, at the same time, being open to new ideas.

Building Shared Vision: Many years ago, I regularly worked for a particular large FMCG company. Every time I climbed the stairs to their reception area, I would burst out laughing. Over the door, bold and unchallengeable, was the statement, "You are now entering a total quality zone". I viewed the statement as a rather crass attempt by senior management to impose shared vision on its employees. The sentiment was 'applied from the outside', rather than growing from within. As such, it provoked amusement amongst visitors and ridicule from employees. This is not the shared vision that Senge talks about. He describes shared vision as, *"the capacity to hold a shared picture of the future we seek to create"* (1990: 9). Often this vision emerges as a by-product of intense involvement in what we do (Kay, 2010: 59-67) and attempting to impose or hothouse a shared vision is likely to prove difficult, if not counter-productive. Although shared vision may be difficult to define, it is nonetheless very tangible. We have all had the experience of entering a company or meeting a group of people who share such a vision. They exude a sense of purpose and passion and generally they are both successful and effective in what they do.

Team Learning: Senge regarded team learning as 'the process of aligning and developing the capacities of the team to create the results its team truly desire' (Senge, 1990: 236). He argues that, when teams work together, not only can there be good results for the organization, but members will grow more rapidly than could have occurred otherwise. He does not assume that all team members necessarily agree with one another.

Some management theorists have criticized Senge's thinking. He has been accused of naivety and few large organizations can be identified that come close to this model of organization, though some would claim it, and others aspire to it. Many have claimed that the Learning Organization is simply not feasible within a capitalist system, in which financial priorities are overwhelming. In my view, this should not stop us trying—just as the knowledge that someday we will die, should not stop us living life to the full!

Complex responsive processes of relating

In the last decade or so, there has been a groundswell of interest in organizational development from the perspective of the complexity sciences or 'complex responsive processes of relating', which presents a radically different way of understanding organizations. Complexity theorist, Professor Ralph Stacey, describes complex responsive processes as:

…the processes of interaction, or relating, which is itself a process of intending, choosing and acting. No-one steps outside to arrange it, operate on it or use it, for there is no simply objectified "it"! (Stacey, 2000: 187)

Stacey is highlighting a key element of complex responsive process, that of *emergence*; in which action and thought are not orchestrated from 'outside' because there is no 'inside' and 'outside' and, as a consequence, there are no boundaries.

Complex responsive processes explain the evolving processes of relating between people who form an organization. This is the basis for a theory, or perspective, of strategy, where human interaction is perpetually constructing the future as the known-unknown, that is, as continuity and potential transformation at the same time. This perspective reflects a theory of transformative causality—a fundamentally paradoxical theory of causality. (Stacey, 2007: 435)

Complex responsive processes theory focuses on 'the living present' (Stacey, 2003), the emergence of ideas, thoughts, feelings and how these develop, shape and are shaped by others in the on-going generation of knowledge. In this respect it has much in common with Qualitative Mind. It provides theory to underpin a qualitative way of working within organizations which is improvisational, creative and adaptive (Keegan, 2008, 2009a, 20011). This way of understanding organizations as fluid, on-going, processes of knowledge generation is sometimes erroneously conflated with current business practices, in which information overload and speed have become business necessities, leading to sloppiness and lack of disciplined thinking. By contrast, organizational life as complex responsive processes of relating demands time, reflexivity and intellectual rigour. The concept of 'emergence' is not a substitute for disciplined thinking (Keegan, 2009a).

Behavioral economics

Ironically, at the same time that quantitative targets were hotting up and spreading throughout government departments and large commercial organizations, behavioral psychologists were making fascinating leaps towards a better understanding of how we make decisions—and the role of conscious and unconscious processes in decision making. Whilst target culture encourages mechanistic approaches to achieving organizational goals, behavioral psychology, (and neuroscience) has been uncovering just how non-linear, complex and (sub) conscious much of our thinking really is.

This school of thought has been labelled 'behavioral economics', although arguably it fits more comfortably within the psychology rather than the economics camp. The author and researcher, Wendy Gordon, has written a concise and authoritative paper on behavioral economics related to qualitative research, which is strongly recommended (Gordon, 2011).

There are a number of conflicting definitions of behavioral economics but, very broadly, it is the use of social, cognitive and emotional factors in understanding the decision making and behaviors of individuals and organizations. This understanding is then adopted as a means of 'nudging' human behavior by framing a decision in a particular way. To take a couple of simple examples: providing condom machines in public toilets can nudge people towards practicing safe sex. Equally siting, 'healthy options' on entry to the school canteen queue has been shown to reduce 'unhealthy' eating.

The theory behind behavioral economics is not new—qualitative researchers working in private and public sector organizations have been using this thinking for decades. However, as behavioral economics has gone mainstream, it has flagged up the weaknesses of our supposedly logical thinking to a wider audience. In doing so, it has thrown into question the efficacy of imposing target culture within organizations. As Nobel Laureate author Daniel Kahneman puts it, "The world makes much less sense than you think. The coherence comes mostly from the way your mind works." (Kahneman, 2011: 58).

Gladwell's theory of 'Thin Slicing'

Malcolm Gladwell (2005), in his best selling book, 'Blink', discusses the concept of the 'adaptive unconscious' (not be confused with the 'unconscious' as described by Sigmund Freud, which is an altogether different concept). The 'adaptive unconscious' makes it possible for us to, say, turn a corner in our car without having to go through elaborate calculations to determine the precise angle of the turn, the speed of the car and its steering radius. We operate on 'automatic pilot', aware of what is happening, but not on a conscious level.

Gladwell describes the 'adaptive unconscious' as that part of our brain that leaps to conclusions. We are all familiar with the sensation of 'knowing something' but not knowing how we know it. We use this ability all the time. We make a judgement on someone within seconds of meeting them and often we are proved right.

Gladwell reports on an experiment to explore this phenomenon. College students were shown three two second videotapes of a teacher they had never met, with the sound turned off. They were then asked to rate the teacher's effectiveness. Apparently, they could do so without difficulty. He discovered that their ratings were essentially the same as evaluations of those same teachers made by their students after a full semester of classes. Watching a silent two second video clip of an unknown teacher was sufficient for students to accurately assess the teacher.

In a different context, we are crossing the road and suddenly see a car heading straight towards us. What do we do? We do not rationally evaluate the risk. Instead, our 'adaptive unconscious' makes an instant evaluation of the risk and our body reacts immediately to avoid danger. Afterwards, we may wonder how we reacted so quickly. If we had depended on our rational brain, we would probably not be here today. Zaltsman goes so far as to say that learning and behavior are dependent on the 95% of brain activity that goes on beneath our conscious awareness (Zaltsman, 2003: 50).

Some psychologists, such as Professor Adrian Furnham (1999: 5-10), would dispute the scale of unconscious activity. Nonetheless, there is general agreement that, although we believe that we make considered and conscious decisions most of the time, this is largely an illusion. From the beating of our hearts to breathing, walking, shrinking from spiders, crossing the road without getting run over, we rely on our bodies to do the work for us, largely without our conscious awareness.

How does our brain do this? Gladwell describes his theory of 'Thin slicing' or 'A little bit of knowledge goes a long way'. Our brain responds to a myriad of details in the situation which our conscious mind is simply not aware of. This detail may arise partly from our intellect, but it will also include input from our emotions, intuition, and bodily reactions. In fact it is a 'whole body', not just a rational, 'from the head' response. We leap to a decision or have a hunch. Our unconscious mind has sifted through the situation in front of us, discarding everything it considers irrelevant whilst homing in on what really matters. Gladwell claims that our unconscious is so good at this that it often delivers a better answer than more deliberate and exhaustive ways of thinking.

In everyday life, we move back and forth between our conscious and unconscious modes of experience, depending on the situation. Neither our rational thinking nor our understanding based on 'thin slicing' is infallible; we have all experienced situations when our 'snap decisions' have proved disastrously wrong. However, by using both modes and emphasizing one over the other according to the situation we can, hopefully, get the best of both worlds.

Our conscious mind takes up so much of our time that it is easy to forget the importance of what happens below conscious awareness in the adaptive unconscious (Wilson, 2002: 17-41). But that is what occurs in organizations when we choose to prioritize quantitative, conscious, 'rational' data and ignore qualitative, below conscious, intuitive data. We get a skewed picture. We miss out on some of the most important and necessary input.

The importance of emotion in thinking

Emotion is one of the most important and undervalued components of the adaptive unconscious. Portuguese neuroscientist Antonio Damasio emphasizes, in his fascinating book, 'The feeling of what happens' (2000) how the brain knows more than the conscious mind reveals. Neuroscience is now confirming what many psychologists, psychotherapists and qualitative researchers have believed for decades; that consciousness is simply the 'tip of the iceberg' and that all sorts of activity which crucially affects our decision making goes on beneath conscious awareness.

What really does challenge our cultural preconceptions, however, is Damasio's assertion that emotion is a necessary component of reasoning. We tend to dismiss emotion as somehow 'lower order'. We talk about *controlling emotion*, *having a rational conversation* and criticise those who are *over-emotional* or *cannot control themselves*. We distrust our emotions because we feel they are unmanageable and we cannot always understand where they come from. Uncontrolled emotion is seen as child-like and unpredictable and we have been taught to distrust what we cannot logically understand or control.

On the other hand, we think of rational, considered thought as being *higher order* brain activity; the most effective way of communicating and an aid to effective decision making, especially within a work context. It is regarded as the evolutionarily peak of our communication abilities.

However, this is simply not true. Damasio's research suggests that having either too much or too little emotion interferes with rational choice. Too much, we can accept, but too little? This seems counter-intuitive in our individualistic, rationally focused culture. It would not seem strange in many Eastern cultures, in which emotion and logic are not set apart as adversaries; the either-or we met earlier.

According to Damasio 'emotion probably assists reasoning, especially when it comes to personal and social matters involving risk and conflict' (Damasio, 2000: 41-2). He suggests that emotion helps with the judgement aspect of decision-making. It provides the emotional intelligence which helps our reason to operate most effectively. It may seem rather paradoxical in our society that, in truth, we cannot make good, rational decisions without emotional input. However, Damasio is at pains to point out that emotion is not a substitute for reason and emotions should not be allowed to reign unchecked. He concludes, "well-targeted and well-deployed emotion seems to be a support system without which the edifice of reason cannot operate properly" (Damasio, 2000: 42).

These different strands in our growing understanding of how our brains 'work' present a very fundamental challenge to the notion of targets as an effective method of evaluating complex patterns of behavior within organizations. Targets are concerned primarily with conscious thinking and generally ignore the below conscious mind. The use of performance targets within organizations as the primary means of understanding or steering human behavior is therefore dangerous territory unless we are very clear about their limitations and effects and, ideally, unless they are coupled with qualitative approaches.

The current socioeconomic climate, coupled with flagging confidence within the UK and European governments and scandals of one shape or form (including target scandals) are forcing a reassessment of the mental models that we unthinkingly adopt when making sense of our world. Linear, reductionist, stop-gap mental models may be comforting in the short term, but they cannot help us to understand the complex, dynamic, unpredictable world that we now inhabit. Maybe the time is right to re-visit a more humanistic, holistic and qualitative way of looking at organizations and, indeed, the world at large.

Fostering Qualitative Mind within large organizations, and developing a climate that rewards employees for building healthy and productive organizations, is essential in order to balance the target culture and build healthy organizations. This is no mean task. Whilst it is relatively easy to develop a learning organization within small companies, it is more difficult to sustain within larger organizations. Hierarchical structures, procedures, rules and regulations are easier to implement and enforce, even when their effects are clearly negative. So how do we start to change?

QUALITATIVE PRODUCTIVITY: A ROUTE TO HEALTHY AND PRODUCTIVE ORGANIZATIONS

What do we mean by Qualitative Productivity?

As employers and employees, I believe we need to be constantly developing *Qualitative Mind*—a mind-set that is creative, curious, improvisational, reflective and reflexive—in order to encourage healthy and productive processes of organization. Of course, this mind-set is useful in the rest of life as well! I use the term, *Qualitative Productivity* (QP) to mean the outcomes that result when we are engaging together as *Qualitative Minds*. It is a similar concept to that described by psychologist Judi Marshall (1999) as, 'Living life as inquiry'. The term encompasses the way in which we interact with each other and how we think, feel and act. At best, QP can act as a counterbalance to targeting mentality within contemporary culture and organizational life.

My understanding of *Qualitative Productivity* is broadly as follows, although inevitably these descriptors sound bland and cannot capture the lived experience I am trying to convey:

- Productivity that is recognized even if it cannot be measured;
- Shared team goals and objectives originating (at least in part) from employees themselves;
- Employee networks that facilitate creative problem solving;
- A sense of 'flow', excitement and enthusiasm;
- Openness, creativity, flexibility in thinking;

- Willingness to listen to people's ideas and build on them;
- Improvisation based on experience, skills, knowledge;
- A sense of belonging, being part of an aligned group.

And more…

It is easy to *talk* about *Qualitative Mind* and about fostering *Qualitative Productivity* (QP) within organizations. Putting these principles into practice is another matter. It is always work in progress, never a finished state. Often it is difficult to sustain the processes of QP because the default setting in most large organizations—the privileged mode—is measurement. Measurements provide comfort and security in a fickle world. There is a perceived, if sometimes misguided, belief that numbers are more 'real', more trustworthy than thoughts, feelings and emotions.

However, we need to remember that most quantitative data that organizations gather and use are a snapshot; inaccurate as soon as they are collected. Although quantitative data is often perceived as 'clean', unambiguous and neat, this is an illusion, especially if it is based on human behavior. It is neat only because the data have been reified; artificially separated from their messy human context. If we want data that reflects 'real-life' more accurately, we have to put up with the messiness—and that means embracing a qualitative perspective. Life is constantly moving on. Qualitative data are fluid and temporal. Change is expected and anticipated, context influences perception and data are contradictory and incomplete. These are strengths not weaknesses because, whilst qualitative data is still an approximation, it is closer to lived experience than most quantitative data.

To illustrate how this can work, I want to outline a project that Campbell Keegan was involved in a couple of years ago. This project, championed by the CEO of the organization, was a very deliberate attempt to develop QP within the target driven environment of the NHS.

Introducing a leadership programme within a Mental Health Trust

The organization was a Mental Health Trust based in London and, like other NHS organizations, it was heavily target bound; staff felt under considerable pressure. Managers described how they felt torn between the dual—and as they saw it—conflicting demands of client care and meeting externally enforced targets. Senior managers acknowledged that targets were useful, but nonetheless believed that meeting these targets distracted them from what they considered to be their 'real' job, which was looking after their clients. They felt 'judged' and pressured by targets, whereas they felt a sense of personal satisfaction when they supported their clients. Consequently, there was on-going tension between these two objectives; client care and meeting targets. In addition, senior managers and clinical staff felt the pressure of silo working, which was exacerbated by the geographical dispersal of different Trust units across the borough. This dispersion often led to duplication of effort and to limited and confused communications. Different units were sometimes wary of one another and so avoided non-essential contact. This, in turn, led to inefficiencies and limited the opportunities for shared learning.

However, the Trust was fortunate in that it had a very enlightened CEO who was keen to implement a leadership programme for senior managers which would address some of these issues. The team was led by a senior manager within the Trust, who specialised in organizational change. In addition there were a number of facilitators; specialists in organizational change work— and me. I had a dual role as a researcher/ facilitator. I monitored and assessed the efficacy of the programme on an on-going basis in order to provide a full report for the NHS, who was funding the programme. In addition, as an organizational change specialist, I participated in the change process itself.

After much discussion between the CEO, the leadership training team and the researcher (me), the exact aims of the programme were hammered out. The leadership programme would focus on increasing networking and shared learning, on providing support for managers who felt isolated and unsupported, and on helping them to deal more effectively with the pressure of externally imposed targets.

The Trust then embarked upon the leadership development programme, which included all senior managerial and clinical staff who were accountable through a director and who had substantial management and leadership responsibilities within the Trust. The programme, which ran for nine months, was designed as a collaborative participative venture with participants developing their own work areas and learning objectives. It drew on recent developments in Complexity Theory to inform the approach, emphasizing the importance of relationship, conversation and interaction in bringing about change (Shaw, 2002; Stacey, 2003). The overall aim of the programme was to develop reflective collaborative networks between senior skilled staff out of which new ideas, initiatives and changes could occur.

Senior managers were keen to reduce work pressure and work more productively. However, many balked at the anticipated time commitments of the leadership programme. They had all been on various leadership training courses over the years, with varying degrees of perceived value, and the prospect of being involved in a programme for almost a year, with 2-3 day stints every month was simply overwhelming. How would they make time to meet their targets? However, the CEO was adamant. If the work culture and patterns of interaction within the Trust were to change, then every senior manager, without exception, needed to attend the leadership programme. With some reluctance, all the managers eventually acquiesced.

Over the course of the programme, we engaged in a wide range of activities which were designed to build up familiarity and trust between managers. A crucial aspect of this was managers establishing networks across the Trust. This allowed managers who were developing similar initiatives to build working relationships, share experiences and act as coworkers, where possible. It increased their sense of autonomy and made their working practices more efficient. Over time, the managers developed and shared strategies. They learnt from one another how and what to prioritize. They developed strategies which enabled them to satisfactorily deal with external targets. They learnt who to go to for help, when needed. They learnt to trust their colleagues on other sites, and work with them in an honest and collaborative way.

Outside speakers introduced new ways of conceptualizing organizations as learning networks, in particular feeding in thinking from the complexity sciences and the role of emergence in organizations. Participants, working in diverse groups, explored how these ideas might be implemented in their particular situation or in specific contexts. Action Learning sets were encouraged. This involved groups of 4-5 participants who met regularly outside the programme days. They jointly worked on real challenges, sharing the knowledge and expertise of the small group of people, combined with skilled questioning, to re-interpret old and familiar concepts and produce fresh ideas.

At one stage, all the members of the Trust Board were invited to join the programme for the day and to talk with the senior managers. The Board members were visibly anxious as they entered the room. As with any hierarchical organization, there was some on-going friction between layers of management. The senior managers were forthright and sought to discuss a number of key issues. It was a tense and difficult time, but the Board Members were open and receptive, in spite of their apprehensions. They acknowledging the issues that the senior managers raised and took on board the grievances expressed. Gradually both sides relaxed and by the end of the day each side claimed that they had gained a better insight into the position of the other. The exchange was deemed to have been a success by all of the participants. It sowed the seeds for networking to extend to and from the Board. Each level became more engaged with levels above and below them; they had a better understanding of the concerns of others and could work jointly with a wider range of staff, when necessary.

The outcomes of the initiative were broad and sometimes unexpected; too lengthy to discuss in detail here. However, to summarize, the large majority of participants believed that they had gained a great deal from the programme, even many who had been very sceptical of 'more management training' before the programme started. However, we were more convinced by their actions than their words. In the large majority of cases, the networks and working groups that were formed during the programme were still operating six months after the programme had finished although there was no requirement for them to continue. Participants had decided to continue the group meetings because they found them useful. Many managers described these as invaluable supports. Although their workloads had not reduced, managers reported feeling more relaxed and more productive. They felt supported. They had colleagues they could call on if need be. Perhaps the most interesting feedback was that they felt better able to meet their targets *as well as* serving their clients well. Quantitative targets and *Qualitative Productivity* had—to a significant extent—been successfully integrated and, arguably, the whole had achieved more than the sum of the parts. Senior managers felt that the organization had become 'healthier' as a result of the programme and their own feelings of confidence, being in control, being supported had filtered down to other staff in the organization. As a result, the CEO planned to implement a similar leadership programme within the next layer of management.

Organizational change as a shift in mind-set

The case study outlined above was very time and energy intensive for both the leadership team and the participants. This isn't always the case. Sometimes organizational change can erupt as a bolt from the blue, through a shift in perspective which re-frames the nature of the organization, its aims or its customer base. This type of re-framing is difficult to anticipate at the start of an organizational intervention, but when it happens it can be dramatic and a breath of fresh air for all who work there.

A couple of years ago Campbell Keegan was invited to work with a large arts organization. The senior managers were unsure about the future direction of the organization and were 'stuck' in terms of developing a strategic plan. "We feel rudderless" explained one of the senior team. This mood was palpable. Various performance targets had been put in place, but it was clear that many staff did not approve of them, or felt that they were addressing the wrong issues. Mostly they were ignored or corrupted in some way.

We talked extensively with staff throughout all levels of the organization. The goals, and indeed the purpose, of the organization seemed very muddled. Gradually it became clear to us that there were two strong but opposing beliefs that permeated the organization and these beliefs lay in the area of responsibility and accountability.

On the one hand, staff believed that their role was to support struggling artists, through funding and through emotional and practical support. On the other hand, there was a rooted belief that the organization was set up to service the general public and therefore providing useful services and value for money to the general public should be their priority. Over years these two issues had become 'rationally invisible' (Shotter, 2003: 60), submerged and inter-twined within the organizational culture, but rarely articulated. They just existed as inchoate beliefs or passing thoughts which could not be reconciled.

In our summing up presentation to the Board and other senior managers, we presented two bald charts, using a simplified version of transactional analysis (TA). One version showed the "Support the artist" model and the relationships, responsibilities and ramifications that resulted from this strategic position, and the second version illustrated the "General public are our main stakeholders" model, along with the ramifications that resulted from this position.

The audience was quiet for a moment as it absorbed the stark decision that the models presented. Then it dawned on them. They had to adopt one strategy or the other. They could not effectively do both. There was then a great burst of energy in the room as small groups convened to discuss the options. The staff felt motivated, and enthused. The choices had become starkly clear in a way that was not obvious before and they felt they could move forward. The bald choice gave them permission to speak the previously unspeakable.

Employees can easily differentiate between targets that help them to perform more effectively and those that have been implemented without any real thought as to their relevance on the ground. The targets that these employees had been set were largely ignored because they did not make sense. Now, by clarifying the options, they could start to create an organization that had direction and that would be healthier for both staff and the people they served.

Integrating qualitative and quantitative approaches

As Peter Senge points out, in order to work well together, members of a group need to be aligned so that their energies (but not necessarily their views) are working in the same direction (Senge, 1990: 233-238). Generally this requires a mixture of self organization by employees and benign leadership. These ideas are not new. Lao Tzu describes this benign leadership as 'the invisible hand'. And this was 500BC!

Similarly, to maximise their effectiveness, quantitative targets and Qualitative Productivity also need to be in alignment, as they became—to a significant degree—in the examples given of the Mental Health Trust and the Arts Organization. In this way each type of understanding can feed off each other and, in doing so, develop a more well-rounded understanding of the issues in hand.

However, in many organizations, quantitative targets and Qualitative Productivity are not actively integrated; the two approaches exist in isolation. In practice, this lack of integration often means that they are experienced by employees as conflicting objectives—either-or rather than both-and. Target setters and those interested in Qualitative Productivity may see themselves in opposition. Where this happens, the 'lowest common denominator' is the likely result, rather than achieving 'more than the sum of the parts'. We see competing forces, rather than Yin-Yang balance.

Peter Senge (1990) describes a number of scenarios to illustrate how employees may work together, effectively or otherwise; the worst case scenario is empowered individuals with low alignment. The result, according to Senge, is invariably chaos. On the other hand, working together, ensuring that both aspects of productivity, qualitative and quantitative, are aligned—and most importantly the outcomes are shared across the organization—is likely to produce a more efficient and a better quality outcome.

SCANNING THE HORIZON

There is a tendency within large organizations to make Gods of rationality; to elevate reason above all other human faculties. At the same time, other forms of knowledge which are acquired through context and practical experience; emotion, intuition and the knowing that is held within our bodies and that we access without even being aware of it, are downplayed, if not ignored altogether. We have a range of ways in which we can 'learn and know' and yet we choose to privilege rationality and devalue other, equally important, ways of knowing that as a species we have developed and honed over millennia. This knowledge bias can be dangerous. In particular, the *belief* that our decisions are purely rational is deeply flawed. Whether we acknowledge it or not, emotion, intuition, bodily knowledge and context are key ingredients in decision making. If we do not appreciate their important contribution, then they will influence us without our knowledge; our decisions will inevitably be biased and perhaps simply wrong. The world-renowned neuroscientist, Antonio Damasio sums this up neatly:

> *The neurological evidence simply suggests that selective absence of emotion is a problem. Well targeted and well-deployed emotion seems to be a support system without which the edifice of reason cannot operate properly. These results and their interpretation calls into question the idea of dismissing emotion as a luxury or a nuisance or a mere evolutionary vestige. They also made it possible to view emotion as an embodiment of the logic of survival.*
> (Damasio, 2000: 42)

Through over-structuring and over–simplifying the way we think, we can create the illusion that we live in a purely rational world—a quantitative world. Then, we reason, if we can accurately define the rules and ensure that the relevant employees play by them, success will be ours for the taking. We can control ourselves, our employees and our world. Once we have convinced ourselves that this is how the world truly is—or should be—then organizational performance targets seem like the ideal driver.

Targets are predicated on the notion of a rational and mechanistic mind that works in a systematic and unquestioning manner to achieve targets which are frequently set by other people. Unfortunately, these targets are not necessarily deemed sensible, appropriate or attainable by the person tasked with achieving them. Those on whom they are imposed, frequently treat them in anything but a 'rational', accepting manner.

Attempting to micro-manage employee responses, setting very specific goals, rewards for meeting targets, performance tables to publicly display employee rankings; this is the model used by the famous behaviorist psychologist, B.F. Skinner to train rats in mazes in the 1950s. It worked quite well on rats, until they worked out the system and turned it to their advantage. The crude assumption behind this 'behaviorist' approach to organizational life and work, is that people will learn what you want them to learn and will behave in the way that you train them to behave. I have tried, in the course of this essay, to debunk this myth. People—and indeed rats—are too smart and too curious to be so easily moulded.

However, while some organizations are seeking the definitive targets, new thinking in management, neuroscience and psychology is advancing in very different directions. There is increasing acceptance of the importance of whole-body knowing, below conscious awareness and the essential role of emotion in effective human functioning. This thinking presents a very fundamental challenge to quantitative-based targets as stand-alone measures for motivating and gauging organizational performance.

Banking on the future

Let us now turn briefly to the bankers, because this is the point where we started this essay and because, arguably, they illustrate the most extreme example of how organizational life can become so target driven that eventually employees can became divorced from reality. How did the banks get themselves into such a state? How did the bonuses keep on rolling as the banks fell to their knees and some senior managers fell on their swords?

I am a psychologist not a banker so I cannot offer a financial explanation, but I can suggest a number of psychological constructs which might help us to explain what happened. We discussed 'group think' earlier in this essay. This mode of thinking typically occurs in decision-making groups which are isolated, culturally or physically. The need for consensus in the group is very strong and this interferes with the ability of the individuals to make realistic assessments of the situation. A consensus decision may therefore be made without critical evaluation. Rather, it is shaped by the shared emotional state of the individuals in the group. In rarefied sectors of the banking world, behaviors that would be rejected in 'normal' society became normalized. To restate the findings of the Levin-Coburn report, *"The (banking) crisis was not a natural disaster, but the result of high risk, complex financial products and….failure of the regulators, the credit ratings agencies and the market itself to rein in the excesses of Wall St."* **Group think** reinforced the niche attitudes and beliefs of an inward looking financial community. **Heightened emotion** which was **contagious** ensured that this wrong-headed thinking spread rapidly throughout the banking industry and became **normalized**. **Greed** and **fear** fanned the flames and **myopia** ensured that most employees could never understand or even see the bigger picture. Instead, they focused on the detail; their own patch, their own dealings, their own targets. They reacted more than acted. It was impossible, it seems, for any one individual to really understand the whole and, as a result, employees from every level of the banking organization could only watch as the banking industry fell like a pack of cards.

Just as 'excessive' rationality is a danger, it is clear that excessive emotion is also dangerous. The bankers may have believed that they were being logical and rational in their responses, but it is clear, at least in retrospect, that they were being anything but rational. They were riding on a wave of emotion masquerading as good business. Rational and emotional forces did not support each other, but instead fought one another. They divided, but neither ruled.

What lessons can we learn that might just help prevent something like this happening again? Well, we need to start with an acknowledgement that we are complex beings that interact within the world in complex ways. If we diminish ourselves, if we deny our true selves by employing simple, 'rational', replicable models of the world in the hope of creating uniformity and easy answers in complex situations, we will achieve a sense of control but ultimately we will fail. We need to try as best we can to utilise all our ways of learning and knowing in an integrated way, with the emphasis being on the quality outcomes we can achieve as organizations comprised of diverse individuals. To do this, we need to re-incorporate qualitative and quantitative mind-sets as we aim for greater organizational health.

What is organizational health?

How would we recognise 'a healthy organization' if we walked into the building? How would the employees work together? How would they recognise and describe quality in what they do? Do employees enjoy working in the organization? How will they change to meet new challenges and demands? How do different levels in the organization work together? What style of language do they use? What degree of individual initiative do they demonstrate? How do they talk about their organization to people outside it? The answers to these questions will, of course, differ by organization and over time, but they are useful indicators of an organization's state of health. The underlying question, however, is 'Why should we care about the health of an organization at all?

Over a number of decades academic studies, as well as anecdotal accounts, have reinforced the same message: Happy, involved employees are more productive and produce better quality work (Layard, 2005: 67-8). A sense of wellbeing makes us more productive and for this reason there is a strong business incentive for employers to be proactive in facilitating employee wellbeing—including their own. But 'happy and healthy' is not enough. In business terms it is a means to an end. We need to develop specific skills. We need to teach employees to challenge their own—and each other's—assumptions, to map elements of complex problems, making the invisible, visible. For instance, empirical evidence shows that high performing teams talk together in distinct ways that lead to greater productivity and profitability, higher levels of customer satisfaction and better relationships with colleagues (Losada & Heaphy, 2004). We need to be more tolerant of experimentation; creating opportunities for serendipity and learning from our mistakes—as Sir Alexander Fleming famously did in his discovery of penicillin. If organizations are to facilitate these skills in their workforce, then individuals need to be able to move seamlessly between qualitative and quantitative mind-sets, so that they can optimize their creative input.

In a benign business environment this fluidity in the workplace might be difficult. In the current climate, it is extremely tough. However much organizations aspire towards an open, creative, self-actualizing working environment in which employees are happy and productive, reality intervenes. The pressures on cost reduction and transformation need to be acknowledged and addressed. The UK lost huge swathes of manufacturing in the late 70s/80s, as these factors were not addressed adequately at the time. Large organizations are wary of repeating history.

Does this mean that we are powerless to influence change? When we think of organizational change we tend to think 'macro'. The reality is that change and constancy are happening, often imperceptibly, all the time, in the day to day interactions of life, for example:

- Complex organizations are in a state of dynamic interaction, both internally and with the communities or markets they serve. A flu epidemic, a riot, a financial crisis can disrupt the best laid plans. We have to accept that, as mere mortals, we have limited abilities to predict, and often we cannot control what will happen. Situations are emergent and require a style of monitoring and reaction which is in accordance with this emergence. Training people to do their job is clearly important. Training people to improvise effectively when there is no 'rule book' is essential. We need to encourage employees how to use their 'whole body' knowing with confidence; to make improvised decisions based on past knowledge, instinct, intuition and current judgement within organizational contexts.
- It has been estimate that up to 80% of organizational time is taken up in conversation. 'Talk' is our key action tool, yet it is often a neglected discipline. Changing the nature of conversations in organizations may be the single most powerful way of bringing about performance breakthroughs. If people change the way they talk, they change the way they think—and vice versa. Teaching people to integrate qualitative and quantitative mind-sets; a both-and rather than an either-or mentality would revolutionise

the workplace. An added bonus is that conversation in itself has been shown to increase happiness.

- In rapidly changing organizational cultures, doing things *to* people (training, structuring, monitoring etc.) is increasingly difficult. It needs resources, time and money. We need to explore ways in which individuals and groups can self-manage to a greater degree. Of course, this risk brings us back to the disaster of self-management that we saw in the banking crisis. Effective self-management needs to be coupled with individual and corporate responsibility, greater personal self-awareness and shared problem solving.

- When economic times are tough, the tendency is to batten down the hatches. A quantitative mind-set, which tends towards the linear, often predominates. Reduced costs, staff losses, high anxiety, lack of trust between management and workers are typical reactions. However 'opening up' rather 'than closing down', i.e., Qualitative Mind, whilst seemingly counter-intuitive, can be more productive. Pooling intellectual resources, collaborative working, cross company teams can all generate different perspectives and solutions.

Summing up

In this essay I have attempted to explore the role of *Qualitative Mind* and *Qualitative Productivity* within organizational life and, in particular, to look at ways in which organizations can become more productive, more focused on quality outcomes and, at the same time, become healthy and happy places for people to work. It is clear that many large organizations have become fixated on targets as a way of measuring performance because targets are simple measures, thought to be easily understood and easily acted upon. As a steer, in simple situations, targets can indeed offer useful guidelines but, as we have seen, in complex organizational environments they are often counterproductive. Targets are crude measures, often not useful—and sometimes extremely destructive. This is particularly true within service organizations such as the NHS, the education system or the Police Service. In these organizations demand cannot be easily anticipated or controlled and there is a constant state of dynamic interaction with the communities they serve. A flu epidemic, a riot, a terrorist attack, bullying in a school can disrupt the best laid plans. These situations are emergent and require a style of monitoring and reaction which is in accordance with this emergence.

While some organizations are seeking the definitive target, new thinking in management, neuroscience and psychology is advancing in very different directions. There is increasing acceptance of the importance of whole-body knowing, below conscious awareness and the essential role of emotion in effective human functioning. This thinking presents a very fundamental challenge to the dominant role of quantitative-based targets as stand-alone measures for gauging organizational performance.

In summary, I have championed the developing concept of *Qualitative Mind* and *Qualitative Productivity* within organizations; mobilizing the considerable skills developed within the field of qualitative inquiry; skills that are sorely needed within many large organizations. I would suggest that *Qualitative Mind* acts as a counterbalance to target driven culture. However, this approach will only work effectively if qualitative and quantitative targets are integrated, with each reinforcing the other in a balanced way. Unless they are conceived and introduced in this way, so that they are aligned in their purpose, they will revert to competing models; either-or rather than both-and.

There is much work to be done in order to in expanding the notion of *Qualitative Mind* and to develop a harmonious tension between quantitative performance targets and Qualitative Productivity. However the benefits of achieving even partial success are likely to impact not only on organizational efficiency and profitability, but also on healthy organizational life. By this I mean organizations that are largely comprised of individuals and groups who feel they are contributing, who are more actively engaged in their work and who are, subjectively, healthy and happier in the workplace—and who produce quality outcomes.

BIBLIOGRAPHY

Blackman, T. (2002). www.radstats.org.ul/no)79/blackman. jtm.

Boyne, G.A. (2002). *Concepts and Indicators of Local Authority Performance: An Evaluation of the Statutory Frameworks in England and Wales,* Public Money & Management, ISSN 0954-0962, 22(2): 17-24.

Brigham, D.L. (1989). *Qualitative Thought as Exemplary Art Education,* Art Education Vole, ISSN 0004-3125, 43(2).

Campbell, R. (2010). *Inside Language,* Proceedings from the Market Research Society, Annual Conference.

Campbell Keegan study (2008). *Living with Recession 1.*

Campbell Keegan study (2011). *Living with Recession 2.*

Chrzanowska, J. (2002). *Interviewing Groups and Individuals in Qualitative Market Research,* ISBN 0761972722.

Chrzanowska, J. (2010). Personal communication.

Damasio, A. (2000). *The Feeling of What Happens,* ISBN 9780099288763.

Earls, M. (2009). *Herd: How to Change Mass Behavior by Harnessing Our True Nature,* ISBN 0470744596.

Farnham, A. *Body Language at Work,* Chartered Institute of Personnel and Development, ISBN 0852927711.

Gordon, W. (2011). *Behavioral economics and qualitative research—a marriage made in heaven?* International Journal of Market Research, ISSN 1470-7853.

Gordon, W. (1999). *Good Thinking: A Guide to Qualitative Research,* ISBN 9781841160306.

Gladwell, M. (2005). *Blink: The Power of Thinking without Thinking,* ISBN 9780141014593.

Griffin, D. (2002). *The Emergence of Leadership,* ISBN 9780415249171.

Janis, I.L., (1982). *Group Think,* ISBN 0395317045.

Kahneman, D. (2011). *Thinking, Fast and Slow,* ISBN 9781846140556.

Kay, J. (2010). *Obliquity,* ISBN 9781846682889.

Keegan, S. (2006). *Emerging from the cocoon of Science,* in The Psychologist Magazine, 19(1), http://www.thepsychologist.org.uk/.

Keegan, S. (2008). *Re-Defining Qualitative Research within a Business Context,* ISBN 9783836474177.

Keegan, S. (2009a). *'Emergent Inquiry': A Practitioners' Reflections on the Development of Qualitative Research,* Qualitative Market Research: an International Journal, ISSN 1352-2752, 12(2): 234-248.

Keegan, S. (2009b). *Qualitative Research: Good Decision Making through Understanding People, Cultures and Markets,* ISBN 9780749454647.

Keegan, S. (2009c). *Co-Creation: The Source of all wisdom or the blind leading the blind,* Proceedings from the annual MRS Conference (Available from WARC).

Keegan, S. (2011). *Qualitative Research as Emergent Inquiry: Reframing Qualitative Practice in Terms of Complex Responsive Processes,* ISBN 978098421651.

Kohn, A. (1993). *Punished by Rewards,* ISBN 0395650283.

Langmaid, R. (2010). "Start listening, stop asking," International Journal of Market Research, ISSN 1470-7853, 52(1): 131-138.

Layard, R. (2005). *Happiness: Lessons from a New Science,* ISBN 1594200394.

Leadbeater, C. (2008). *We-Think,* ISBN 1861978370.

Leapman, B. (2007). The Telegraph, 8th July, *'Police 'target forms, not criminality'.*

Lehrer, J. (2009). *How we Decide,* ISBN 0618620111.

Losada, M. and Heaphy, E. (2004). *The role of positivity and connectivity in the performance of business teams: A nonlinear dynamics model,* American Behavioral Scientist, ISSN 0002-7642, 47(6): 740-765.

Marshall, J. (1999). "Living Life as Inquiry", *Systemic Practice and Action Research,* ISSN 1094-429X, 12: 155-171.

Mead, G.H. (1962). *Mind, Self & Society: From the Standpoint of a Social Behaviorist,* ISBN 9780226516684.

Moscovici, S. and Zavalloni, M. (1969). "The group as a polarizer of attitudes," *Journal of personality and Social Psychology*, ISSN 0022-3514, 12: 125-135.

O'Neil, O. (2002). *A Question of Trust*, Reith Lecture, No 2.

Quick, J.C. and Quick J.D. (2004). "Healthy, happy, productive at work: A leadership challenge," Organizational Dynamics, ISSN 0090-2616, 33(4): 329-337.

Rose, D. (2012). "400 deaths linked to appalling care at Trust," The Times, 25th.

Rogers, C. (1961). *On Becoming a Person: A Therapist's View of Psychotherapy*, ISBN 039575531X.

Schon, D.A. (1983). *The Reflective Practitioner: How Professionals Think in Action*, ISBN 9780465068784.

Seel, R. (2000). "Culture and complexity: New insights on organizational change," *Organizations and People*, ISSN 1350-6269, 7(2): 82-84.

Senge, P.M. (1990). *The Fifth Discipline*, ISBN 9780712656870.

Shaw, P. (2002). *Changing Conversations in Organizations: A Complexity Approach to Change*, ISBN 9780415249140.

Shotter, J. (1993). *Conversational Realities: Constructing Life through Language*, ISBN 9780803989337.

Stacey, R. (1996). *Complexity and Creativity in Organizations*, ISBN 1881052893.

Stacey. R. (2001). *Complex Responsive Processes in Organizations*, ISBN 0415249198.

Stacey, R. (2003). *Complexity and Group Processes: A radically social understanding of individuals*, ISBN 0415249198.

Stacey, R. (2007). *Strategic Management and Organizational Dynamics: The Challenge of Complexity to Ways of Thinking about organizations*, ISBN 0273613758.

Thome, D.C. (2009). *Mid Staffordshire NHS Foundation Trust: A Review of lessons learnt for commissioners and performance managers following the Healthcare Commission Investigation 24th April 2009.*

Watts, A. (1954). *The Wisdom of Insecurity*, ISBN 0307741202.

Watts, A. (1969). *The Book on the Taboo Against Knowing Who You Are,* ISBN 9780679723004.

Wilson, T.D. (2002). *Strangers to Ourselves: Discovering the Adaptive Unconscious,* ISBN 0674013824.

Williams, M.D. (2010). "Exploring 'system resilience' in the context of a hospital bed crisis," in M. Anderson (ed.), *Contemporary Ergonomics and Human Factors*, ISBN 04155844469.

Wheatley, M. (2002). *Leadership and the New Science: Discovering Order in a Chaotic World,* ISBN 15767511198.

Zaltman, G. (2003). *How Customers Think: Essential Insights into the Mind of the Market,* ISBN 1578518261.